# RESILIENCE
## ADAPTATION
### SUSTAINABILITY

# RESILIENCE
# ADAPTATION
# SUSTAINABILITY

## ROBERT RIDDELL

Published by Robert Riddell
Auckland, New Zealand

eBook available at www.resiliencebook.net

This edition published 2014

The right of Robert Riddell to be identified
as the author of this work in terms of section 96 of the Copyright Act 1994
is hereby asserted.

Designed, typeset and produced by Mary Egan Publishing

MARY EGAN
PUBLISHING

www.maryegan.co.nz
Printed in China

ISBN 978-0-473-29245-4

# CONTENTS

*Resilience Adaptation Sustainability* targets a young adult readership concerned for their children's and grandchildren's future.

A quarter of the book's content is devoted to factual referencing and support reasoning — from Appendix A through to the last endnote.

# BACKGROUND

The ambition for this book and the companion low-cost e-book is that it motivates younger adults, those who 'own' and will 'be in charge' of the future.

My writing has its origin in a variegated upbringing, fortuitous schooling, project challenges, academic wrangling, relationship enrichment, and the advent of children and grandchildren.* Relevant experience came from significant commissions in Oceania, West Africa, Kashmir and Amazonia. Relative enlightenment came through from the social hurt and environmental degradation roiling away in the OECD and OPEC nations of self-believed sophistication. Over time, personal reflection cultivated an itch in the back of my mind, which led to a synthesis (Part 'A') and the production of an advisory (Part 'B') for sustaining our children's inheritance.

Recent guidance and high-end support from Barry Pearce of Cambridge times, Robert Collin and Robin Morris Collin at the University of Willamette, and Jacqueline Margetts and Rod Barnett of Auburn University, is warmly appreciated. Daily life for Heather and myself is invigorated by positive goodwill from the children and staff at Helensville Montessori, and through our grandchildren — Phoebe, Grace, Bonny, Oliver, Danny, David and Andrew. I am thankful and grateful to be part of Marunui Conservation; the fellowship group and bush-and-bird backdrop for an off-the-grid cabin where the first draft was written.

Although self-published and available as an eBook, the script is neither self-edited or self-formatted; the credit here, and my heartfelt thanks, go to Bernie Ranum and the team at Mary Egan Publishing.

Bob Riddell
Helensville, New Zealand
catch21@ihug.co.nz

* Selections from the *Salient Happenings* in my life can be viewed at:
www.resiliencebook.net

# INTRODUCTION
## FUTURE PROOFING

**This writing brings 'sustainability' the ideal, and 'adaptation' the action, together.**
The environmental change that is upon us arises from a chilling gradualism; ppm-by-ppm (parts per million) carbon gas increase, degree-by-degree temperature increase, centimetre-by-centimetre ice melt, sea level rise millimetre-by-millimetre. These erosions and accretions are now undeniable, the new normal. Melted permafrost will not refreeze, savannah extensions will not reforest, sea level rises will not recede, exterminated species will not re-evolve, and displaced populations will not repatriate. We have seen this coming and have a fairly clear idea where it is all going.

Shelf-upon-shelf of books and journals call on us to lessen the ugly impact of biospheric change and ecological shrinkage by consciously lowering our personal levels of resource consumption and waste discard; the rhetoric for adjustment lagging behind the advance of climate change.

Within the remainder of this century cumulative degradation of the biosphere will be of such magnitude, and become a matter of such immediacy, that adaptation will shift from a premise to a fact. This situates humankind 'between a rock and a hard place'. That is the 'rock' horror of consumption and discard laying waste to our living space, and the 'hard place' sacrifices associated with sustaining a liveable future.

What is essentially a social challenge comes down to slowing, arresting and reversing the violation of our biosphere. It is a course of action worthy of, but currently beyond, our collective intelligence. What it calls for is a move from hedonism to contentment, from exploitation to conservation, from self-interest to common good, from outside dependency to local sufficiency; 'less' working through eventually as 'more'. The difficulty with these truisms is that most of us, myself included, doubt whether such 'aspirational' recommendations can be effective while the counter-intuitive workings of market fundamentalism and print money supply prevail.

Along with expanding consumption, institutional decline and conflicting political power our economic system, drugged-up on electronic money, sucks fossil fuels out of the ground and discards the harmful residues 'free' into the biosphere. The overwhelming frightener here is increased entropy, the disaggregation and dispersal (but not the disappearance) of fossil carbons into unusable heat, waste and contaminants.

Because this process is at an early stage of overload and misunderstanding it is not rational to expect a fulsome uptake of the soft-pathway advocacy any time soon. But, eventually, humanly driven biospheric change will induce severe systemic shock. At that future date societies and communities will hopefully, surely, look for ways to adapt.

---

**Part A evokes an understanding of the biospheric 'Limits of Resilience'**
**Part B provokes the 'Actions & Adaptations' for a sustainable future**

---

The surround-objective is to assist people with a wrong-right conscience to better comprehend the limits of biospheric resilience and the function of adaptation; and to tender a roadmap (in Part B) for those who seek furtherance of a sustainable culture as our children's legitimate inheritance.

---

**There exists a nexus of financial and corporate management which governments have become almost powerless to change; not consciously, yet actually, in line with Ayn Rand's odious "...virtue of egotistical selfishness", complicated by John Gray's discomforting notion of "...a world without meaning".**

---

The story of damage to the biosphere over the last century identifies the enemy. It is us, and those we entrust with political power. Thereby, the logical way to work out what is wrong with the environment is to first work out what is wrong with our institutions and ourselves. We can start with the ideology of growth, the culpability that fosters consumer profligacy and induces apathy around waste discard. What is now apparent is that print money and faux credit feeds the torpor that produces the waste that is killing the environment.

Humanly driven biospheric change induces 'future shock' and at some point — again, within the remainder of this century — this will manifest as a massive global shift in environmental stasis, jolting humanity away from descending deeper and deeper into dystopia.

## IN TERMS OF MORAL INTEGRITY (CONSCIENTIOUSNESS):

Deliver to the ideal of habitable wholesomeness, economic stability, responsible 'green consensus' politics, socially sensitive fertility equilibrium, and compassionate environmental justice. Not only for ourselves. Mainly for our children's children.

## IN TERMS OF MORAL ACTION (COMMONSENSE):

Adapt to a socially-proactive, institutionally prescriptive and baseline-pragmatic resource management advisory. Put an end to made-up money; embed fiscal stability. Sensitively edge fertility rates below death rates. Eke fossil carbons out of the ground slowly, use them efficiently, tax them heavily, and punish their misuse mightily. Obtain bulk energy supplies from free-flow sources. Curtail nitrogen fixing. Extend permanent forest cover. Avoid, reduce and recycle waste.

# PART A
## LIMITS OF RESILIENCE

"A society that destroys the environment
that sustains it will fail … what is needed is
a change of heart."

— ANNE SALMOND

**Part A: Limits of Resilience** examines patterns of change and the limits of global resilience to change. The first two chapters explore human behaviour and settlement expansion. Chapter 3 traces consumer growth, planetary limitations, and the environmental consequences of waste discard. Chapter 4 identifies the emergence of environmental awareness. Chapter 5 provides a planetary overview of population growth, Chapter 6 technological advance, and Chapter 7 print money supply.

Appendix B provides 'Warnings from Oceania'.

---

Readers with an understanding of the systemic causes of climate change, or badgered replete and fed-up with the harrowing bleakness of it all, may choose to go direct to the prescriptive **Part B: Actions & Adaptations**.

---

# 1

# THREE COMMUNITIES
# TWO FAILURES

---

---

A few hours before Neil Armstrong took that momentous step onto
the moon's surface my daughters, then infants, squeezed between their
mother and I on the family scooter. We tuk-tuked along pedestrian trails
in the cooling late afternoon to explore an archaeological dig, work
finished for the day, at a tree-shaded spot by the side of a stream running
through the vast Kumasi University Campus in Equatorial Ghana.

Here workers were unearthing the traces of a Stone Age technology
left behind well over 20,000 years previously. The place was known
locally as The Grindery; and to this precise spot our early forebears were
attracted across the African Savannah, carrying stone blanks for shaping
and polishing. To them this was both a place of technological satisfaction
and some technological failure; evident from the fridge-sized pile of
split shards, broken and discarded while they were being ground. Stone
Age people visited The Grindery over a long period of time, improving
their lives, leaving behind an environment virtually undamaged by their
occupation.[1]

Some months after visiting The Grindery we motored around Easter
Island exploring the commanding statues. The story of how these
enormous stone figures were put in place, by whom, when and for what
purpose, had not been resolved at the time of our visit; they were a marvel

and a mystery. Revelations, when they came to light, were dramatic, yet logical. Stone-working people, ancestors of the Polynesians still living on the island, had built these sentinels for a deity about 20,000 years after, yet not much different technologically, from their stone-working forerunners in Africa.

The process cost the Easter Islanders every tree and induced massive soil erosion; bringing about the end, for them, of wholesome existence. They began to fight among themselves, and by the time of their European rediscovery in the 17th century their population had begun to decline. It became futile to stay; yet they had no 'elsewhere' to relocate to. The scattered shadow population slumped into subsistence. They were without an identifiable enemy, yet they were defeated. Easter Islanders brought about social decline through resource destruction of their own volition.[2]

Back in Ghana, when necessary, we took the Volkswagen Beetle for repairs at a place which was and still is a technological maelstrom — a community known as Magazine — half a square mile of organised chaos, on the northern perimeter of Kumasi. Here vehicles caught in the crossfire of our World's worst traffic accident statistics were delivered for repair or reassembly. Rear-ended taxis were matched to the front-ended wrecks of private cars; and Magazine operators spoke gleefully of two vehicles, towed-in from the same smash, emerging a few days later as one, with some spares left over.

Magazine is remembered as a cacophony of hacking-apart, putting-together and painting-over; testimony to the adaptive capability of modern Ghanaians using technologies learned in their lifetime, applied to materials unknown to their grandparents, which will be denied or in short supply to their grandchildren.

Every human grouping seeks to advance by leveraging advantage from other human groupings, and from the available resources and commons. They, and now we, have been accomplished at this. Homo sapiens have evolved into a dominant super-predator 'successfully' plumbing the deepest ocean, drilling the thickest ice cap, and monopolising the entire planet. Added to which, for the first time in earthling history, more than half this growth-dependent population resides in urban places where we are largely reliant on ex-urban food, fibre and energy supply.[3]

Returning to those snapshots: The three situations reviewed at 20,000, 1,000 and 100-year stages.

**First:** The stone-age visitors to The Grindery engaged local resources and left their place of work untroubled; for they never depleted the flint supply, nor did they wear out the sandstone deposit; living technologically little changed over centuries of time.

**Second:** After landfall the Easter Islanders set about exploiting what was available to them; depleting within a millennium an essential life-enhancing resource, timber; the last felled tree downgrading at a stroke the quality of their living and inducing a population decline.

**Third:** Modern-day operators at Magazine can expect reliable access to the machines and the fuels to drive them for little more than one-tenth of a millennium. With this particular and recent situation all the resources consumed are from other lands; the imported plastics, metals, synthetic fibres and other carbon derivatives and fuels.

---

With less than half of the World's liquid oil resources depleted thus far, there may now remain only a few decades left as we all plummet into Martin Rees's *Final Century of Civilization*, reducing our home to an undone biosphere.[4]

---

The Grindery people's continuum (timeless occupancy, slow technological change and minor environmental impact over several thousand years) contrasts with the decline of the Easter Islanders (resource depletion and social decay within a millennium). The modern Magazine people's future is more conjectural.

Relative to longer-gone communities (the Sumerian city-state located on the plain between the Euphrates and the Tigris a mere five-thousand years ago, the Mesoamerican Maya and the people of Fatepur Sikri in Northern India) it is the closer-in-time ecocide of the Easter Islanders which is most fascinating because of the way they slid into social chaos as a causal consequence of their own undoing, destruction of their habitat; something salutary for 21st century humankind to hold a mirror to.[5]

A common thread is that those ancient groups, unlike the Easter Islanders, dispersed in the way the more recent Dust Bowl drifters resettled from Kansas to California. The problem for the Easter Island people was that they were stuck at home unable to relocate. They were like a socio-environmental research topic awaiting enquiry, analysis, understanding and explanation.[6]

The presence of humankind on planet earth and the circumstance of ecological variety and human existence is an 'extraordinary wonderment' which includes family, beauty, rhythm, reason, argument, art, poetry, and a collection of disparate beliefs. However successful we consider the current arrangements to be, the situation early in the 21st century is that human society has, over the previous hundred years, consumed mineral carbon resources at rates of carbon gas effusion that exceeds the absorptive capability of the biosphere.[7]

## EASTER ISLAND — EARTH ISLAND

The Easter Island experience is frequently cited in relation to Earth Island; both being places with nowhere to retreat to. Most telling, after the tipping point of soil-loss, water-shortage and tree-denudation, was socio-religious and institutional decay. Community mayhem and population decline was then swift and cruel; a hint of what Earth Island – Easter Island on a global scale – could be in for?

Yet the Easter Island example also predicates a positive. Were we drawn to learn from that story, and to form an appreciation of why it is in the interests of our species to provide a stable environment for our grandchildren, then we would have attained the capacity to prolong a wholesome lifestyle beyond hedonism and consumerism.

Contemplate the Easter Island scenario as an Earth Island indicator. And by way of extra reinforcement retain Al Gore's use, in *An Inconvenient Truth*, of that all-alone image of our planet photographed from a space station hurtling away from the solar system – a single blue dot set against a black soulless void!

The populations of the forty-three OECD and OPEC nations revel in hi-tech living; inducing a preoccupation for mobility and gadgetry which blinkers them against ecological overshoot, and sleepwalks us all into environmental disaster. This one-fifth of the global population holds to the belief that monetary leverage is the way to resolve problems, a posture not available to the other four-fifths of the population that is poor. Mass wealth and throwaway waste is as recent as the invention of 'made up' money supply.[8]

> Never prior to this past century has irreversible global destablisa-
> tion been created so extensively and lastingly by human agency;
> nor have irreplaceable resources previously disappeared so rapid-
> ly on a global scale. And never before has so much toxic material,
> which nature cannot absorb, been jettisoned into the biosphere.[9]

One remedy involves lowering the current rate of resource consumption in developed countries to the Bangladesh norm. Paradoxically, in terms of sustainable ideals, this and other cash-poor populations lead a life of near-sustainable perfection. A 'raising' of global living standards to North American levels is of course impossible; for the simple reason that the 'footprint' specialists (Wackernagel and Rees) calculate the current global population, living at Stateside rates of consumption, would require five more planets from which to obtain the requisite resources.[10]

Throughout all of passing time there exists an ever-changing ecological balance. There is no ethereal intelligence or soul at work here.

Simply put, if super-dominance is attained by one specie, adjustments will occur within its habitat, and there will be consequences for all other species. For our post-Millennium biosphere, massive overcrowding and turbocharged wealth has led to one or a combination of harrowing adjustments and lessons: climate change, sea level rise, forest cover decline, disease, pandemics, viral plague, species loss, famine, genocide, revolution, civil war, accidental and deliberate nuclear holocaust.[11]

In a seriously blinkered way we assess our wellbeing from the standpoint of self-wealth, and customise our brief personal lifetimes as the time-scale for working out our own, and our children's, survival. Growth fixation and wealth garnering by individuals corporations and nation states can be identified as the main generator of humanly induced climate change; the greatest market failure ever. Fossil carbon consumption, population overgrowth, and free-to-air-and-water waste discard, accumulate to degrade the environment our children and grandchildren inherit.

Confusing, from an avalanche of writing, is the matter of what needs to be done, and how it can be done? That challenge is taken up in the prescriptive *Part B: 'Actions & Adaptations'*. The critical question now is the matter of 'why?' — 'why should we' the most technically advanced generation ever, be concerned for a future beyond our own lifetimes? After all, it can be countered, it is a future we will not be part of; except via our children and grandchildren![12]

Climatic seasons set the temporal framework within which pre-20th century humankind coped with excesses and shortfalls. Now wealth and technology enables us to respond whimsically to consumer instincts regardless of the seasons — our fad foods, single-generation residences, electronic toys, and prosthetic automobiles. Complementary to this dramatised point is Benjamin Barber's detection of sinister corporate manipulations which infantilise the democratic process and condense adults into 'kidult' consumers. His imputation being against damaging practices we morally and intellectually deplore.[13]

War famine and disease are ugly and stupid ways to arrive at a globally balanced population. During the last century, one hundred million people (one-twentieth of the global starter population at year 1900) died in and as a result of wars. These decimations were accompanied or followed by ever-increasing bouts of pandemic disease, economic depression, violent crime, genocide and ethnic cleansing. These are examples of cruel and heartless dieback processes correcting human imbalance within the biosphere, evident today as the underlying drought-driven basis to conflict in the Sahel, Syria and the West Bank.[14]

If we lack the motive and fail to adopt the means to reduce our fossil fuel consumption and carbon gas discard then the climatic, biospheric, monetary and social upheaval will be unstoppable and hurtful.[15]

We seem largely to know 'what' has to be done; we simply do not know 'how' to do it. And, more pertinently, we have yet to fully and factually appreciate 'why' we should make the adjustment. Humankind is caught-up in a 21st century bind of its own creation, a 'Catch-21' parody of Joseph Heller's *Catch-22*.[16]

## FURTHER READING

Diamond, *Collapse: How Societies Choose to Fail or Succeed*.
Sahlin, *Stone Age Economics*.
Ponting, *Green History of the World*.
Wright, *Short History of Progress*.
Mumford, *Technics and Civilization*.

# 2

# HUMANITY
# AND THE BIOSPHERE

---

The Drivers of Environmental Change and Ecological Destruction | Fiscal Leverage | Carbon-Fuelled Capitalism | Principles of Sustainability | The International Panel on Climate Change

---

The twenty-thousand year (Stone Age Africa) and one-thousand year (Easter Island) perspectives remind us of the different ways those communities — African outward dispersal, and Easter Island inward decay — came to evolve their patterns of living. The lessons are salutary; yet it is the third perspective, within the one-hundred year span of contemporary Magazine in modern-day Ghana, which mirrors the situation in which wealthier nations are embroiled today.

---

**THREE OBSERVATIONS:**

FIRST: The oil-fired modern industrial age, paralleled at Magazine, will be short, gone in a two-century blink.

SECOND: The worst-case global aftermath will be environmental exhaustion; a warmer biosphere with increased carbon dioxide gas in the atmosphere and soluble nitrates in the hydrosphere.

THIRD: Despite these negatives, a positive for the electronic age is that humankind has adapted previously to climatic adversity.[17]

---

Naysayers, unencumbered by facts, deny the dangers of human-induced climate change despite widespread ratification of reports put out regularly by the Intergovernmental Panel on Climate Change (IPCC). To logicians the journey of climate decline will continue, unless fossil carbon consumption and the destruction of natural forest cover are reduced drastically. The daily news — floods, droughts, hurricanes, oil slicks, crimes, and disease, all on an escalating scale — confirms that lifestyle change is with us. These are shifts aligned with degradation of the biosphere as a whole; ecosystem decay, lessened biodiversity, resource depletion, sea level rise, and toxins accumulating in the atmosphere, on land, and in the hydrosphere.[18]

At year 1800 looking to 1900, and at year 1900 looking to 2000, the future would have appeared desirable and improving; and were you in any doubt about the role and destiny for humankind, your religious and cultural belief helped, as science did not then have reasonable and sufficient explanations for the mystery of it all. For most of us the science is now much more certain about the global limits of resilience. Now, looking out to 2100 from our current standpoint is not encouraging. We are at a complex fork in human history. It is the place in time and fact where we have a brief opportunity to decide whether to further extrapolate our recent pattern of ever-continuing growth, or to string together a sustainable survival strategy.

Language-gifted humans, upright and opposing-thumbed for tens of millennia, have become species eclipsing; acquiring the habit, along the way, of falling in with prop-up belief. Like the doom prophesies surrounding the Millennium Event and the 1980's European acid rain scare, some of these fears are misplaced.

Human dominance of the global domain is assured, all the way 'back' to an altered future; potentially a much-altered human grouping and a much less wholesome biosphere.[19]

In sweeping terms Homo sapiens has made no evolutionary mistake. Dealt walk-talking, orbital-thumbing and a foreseeing brainpower we moved to eclipse all other dumb and subservient species. How far we have 'evolved' and how 'consumerist' our orientation has become, is examined in the Appendix A construct 'Consumer Behaviours Conceptualised' centred on a four-part social pathology — Emotions, Thoughts, Sensations and Behaviours.

Climate change during the current part of the Industrial-Electronic Age is now explicit; change being the outcome, consumption of naturally stored 'dense and convenient' energy carbons the cause. The consequential accumulation of greenhouse gases in the atmosphere, and nitrogen in the hydrosphere, induces adverse effects. What is also now clear is that the increase in global temperature and changes to climate results from the carbon gas emissions and fixed nitrogen runoffs that the biosphere cannot absorb; and this is exacerbated by population expansion, wealth accretion and technological advance.

## IPCC: The UN Intergovernmental Panel On Climate Change.

The key word in the acronym is the verb, inferring with some verification (2010 Climategate), that collators were not interested in reports which showed nil or negative change; and that they sought-out reports which indicated temperature increases. The Fifth Assessment Report (2013) involved 209 lead authors.

The first (2007) report was well received, though with some doubting of facts, accusations of data smoothing, over-egged speculation, and bias toward consensus. The 2013 IPCC 'Summary for Policy Makers' conveys, as emerging fact, and as extremely likely, that humanly induced global temperature increase, trending to $+4^{\circ}C$ above Millennium levels, will induce a slow irreversible and persistent melting of the Greenland and Antarctica ice fields, in association with a gradual expansion of the slowly warming oceanic water body; bringing about a 60–100 cm rise in sea level this century. And for this to also cogenerate other 'extreme events' during the 21st century; glacier retreats, a thawing of the Arctic Tundra (involving the release of millions of tonnes of locked-up methane), a 10% loss of food production, a rapid increase in disease, and a 20% loss of species.

Supplementary to 2007 and 2013 IPCC reportage are the 2012 evidential results from the Berkeley Earth Surface Temperature (BEST) project showing a $1^{\circ}C$ increased level of land temperatures over the previous fifty years. Also within the UN is the Framework Convention Convention on Climate Change, which hosts an annual Conference of Parties.

Over recent decades, until the recent economic surge from China and India, a lid has been kept on the number of high-end energy consumers; constrained in effect by poverty. Now, with so many of the previously poor empowered by new fortunes, the levels of energy demand and consumption, along with the consequential adverse climatic effects, are all on the increase.[20]

Poverty is a proportional matter. It has always been so. Thus four billion living in poverty in the early 21st century is proportionally worse, in humanitarian terms, than one billion poor in the early 20th century; provided poverty 'then' is being adequately compared to poverty 'now'. Another point is that all wealthy nations harbour poverty pockets, all modestly wealthy nations have rich-and-poor strata, and all poor nations include enclaves of wealth.[21]

Only one-fifth of the global population is isolated from poverty. Of the other four-fifths, comprising the poor, about half remain 'stone age poor' possessing only the most meagre life-support trappings; a situation not much altered during the course of the Industrial-Electronic Age. Important to note here is the fact that individually an average poor person represents, at most, a one-fifth environmental threat relative to an average rich person.[22]

## ON POVERTY

The 'too poor' elevate our concern; and the 'too rich' dominate climate change.

A base line definition of 'too poor' put out around 1990 was a dollar a day minimum, the Millennium Goal One. Two important distinctions here, which go some way to highlight the injustice of the concept, are the difference between a wealthy individual's 'impulse dollar' spent on a can of coke, and a poor person's 'cautious dollar' spent on family rice; and the fact that both rich and poor pay the same unit price for their electricity and gasoline.

There is no way wealth adjustment could arrest climate change, or disappear poverty, or do both. Neither wealth distribution nor growth has overcome poverty; and neither have had an appreciable effect on climate change.

The *per capita* contribution by poor people to climate change is much less than the rate of the wealthy; yet it has to be noted that the five billion most poor aggregates about the same mass of overall carbon emission equivalence (notably through deforestation) as the one billion more wealthy; thereby ensuring that the climate change problem is also a poor people's concern.

Of the capitalist system Karl Marx (paraphrasing loosely) declaimed in pique when the 19th century Gold Standard first came under threat, that eager feudalists and rampant capitalists will now be able to produce something from nothing. Indeed the deluge of money now econometrically conjured, proved too tempting to resist for corporate bankers and the captains of industry. The resultant problem demands both a reduction in the mass of consumers and in the levels of consumption; calling for a well-regulated, steady state monetary system; based, possibly, on a 'virtual' Gold Standard.[23]

## FOSSIL CARBONS: STRANDED ASSETS

The Carbon Tracker and Grantham Research Institute 2013 report *Unburnable Carbon* is mainly concerned with financial risk exposure for the copious trillions of investment dollars geared to the exploitation of difficult-to-access carbon reserves (See Endnote 25). Vast fossil carbon stocks exist, but all the easy-access stuff has now been taken. Clearly, increased mining costs and higher purchase prices will follow, inducing persistent global recession.

Unburnable Carbon envisages the bursting of a 'carbon investment bubble' on a scale that will dwarf the 2008 'sub-prime investment bubble' the grandmother of all previous boom and bust episodes. The report's tacit advice is for fossil fuel investors to divert their attention to renewable energy manufacture and the installation of renewable energy plant – provided, one has to presume, there is a market in which to divest.

Back to earth, in fact in the earth. Here is stored the remaining, less accessible, fossil carbon deposits which humankind extracts in order to mobilise, accessorise and extend the modern living arrangement. Moderated oil and gas extraction alone would not switch off climate change, nor arrest carbon consumption, nor achieve climate stability. This is so because of the enormous quantities of brown coal and oil shale (oil rock!) available for future exploitation, including hydro-fracking (fracking) of tar sands, all of which require high levels of energy input relative to energy capture. The ever increasing output of brown coal, the main $CO_2$ generator, continues to feed power stations, commissioned at the rate of one a week in Asia; America providing the exemplar consumer format, Australia being both a major brown coal user and supplier.[24]

Falloff in the extraction of liquid oil (after Peak Oil) and the non-recovery of production volumes, coupled to an increased uptake of dirty carbons, will contribute further to socio-economic disorder and further exacerbate climate change. An individual-by-individual complication is that oil is the densest form of conveniently available, useable and affordable energy; which clashes with the reality that mineral carbon denial is seen to be the key to climate control and a wholesome future. This relationship is supported by a succession of recent fiscal recessions — five or six in all — which correlates wealth erosion as leading to reduced oil extraction, followed by wealth resurgence leading to increased oil production-consumption seeking to get back to a state of reprofiting.[25]

The specific timing of peak oil extraction and supply will be a historical reference point, but one of little discernible consumer impact at the time. With growth-on-growth policies (Schumpeter's 'creative destruction') in place, the really serious economic and environmental hurt will occur later. Following a period of global economic downturn associated with carbon gas overburdening, nations will be prone to slide country-by-country into economic-ecologic crises, constrained consumption, stalled industrial activity, diminished food supply and curtailed goods and people movement.

And in all this, what about biofuels? Apart from the unfavourable energy-in to energy-out relationship, poor densely settled people need the land taken over for producing ethanol and biodiesel (the 'joke' liquid carbons) to grow food crops to maintain their food security. At base gasoline production from food crops is a political ruse to reduce unemployment, despite the known inefficiencies and the environmental futility surrounding ethanol production. By excluding the use of food-crop land for sustenance, the jurisdictions which promote crop biofuels must expect the 'fields for fuel' to be eventually set ablaze.[26]

European, Settler Society and other OECD populations are bound into the house-plot-car-accessories syndrome; mainly suburban dwellers addicted to an 'all resources imported, all wastes exported' dependency; every material desire wheeled-up to the door, every particle of waste whipped away.

This style of living is also linked to the manufacture of automobiles and the roads to run them on, in and above which are installed pipe and wire services. Worst off as a consequence of energy curtailment will be automobile dependent 'sunbelt' populations; retirement tracts in near-desert environments where the out-of-doors ambience is mechanically adjusted from being either uncomfortably hot or uncomfortably chilly for up to twenty hours each day.[27]

Suburbia is a long way off the mark in terms of long-term viability

— sustainability. Suburban dwellers on larger plots in rain watered environments, particularly in broadacre (US) and lifestyle (Australasian) situations, will fare better, although they will be mobility-constrained in terms of conventional vehicle use on account of the high cost and unavailability of gasoline.[28]

## SUSTAINABILITY

Definitions of sustainability (striving for 'intergenerational equity') abound, it being easier to identify what is not sustainable.

The most quoted definition of 'sustainable' comes from Brundtland's 1987 publication *Our Common Future*: being to 'meet the needs of the present without compromising the ability of future generations to meet their own needs'.

My preference: 'using, conserving and enhancing community and global resources so that the ecological processes, on which life depends, are maintained indefinitely; leaving the quality of the environment unimpaired'. Sustainability cannot be achieved; it's an objective to strive for within the context and framework of Environmental Justice. 'Sustainable Development' is an oxymoron.

An up-to-date bibliographical resource is the 3-volume *Encyclopaedia of Sustainability* (2010) compiled by husband and wife team Robin Morris Collin and Robert William Collin.

It can be assured rather than assumed that for higher density urban properties there will be service cuts and a falloff with maintenance; this will include restrictions to the inward supply of water and power from distant sources resulting in the necessity of waste disposal on site.

Adaptations can be made; primarily to resource capture through rainwater harvesting, energy capture through hydro, solar and wind generation, solar heat storage, and the likes of kitchen gardening. Boring, this sustainable lifestyle need not be, as the wire-less electronic information age can continue solar-powered, wind-powered and hydro-powered; and the process of living from local resources within redefined communities can be not only a challenge but a stimulus. Adaptive skill, humour in dour situations, and community togetherness will come into play as useful traits, as has always been the case during times of adversity. In point of fact, as argued by Rebecca Solnit using North American histories, adversity nurtures co-operation, the formation of community, and happiness; our default setting in times of challenge.[29]

The time has come to think sharp, get wise, slow down, and act smart. As individuals, and group-on-group, we are a competitive species. But also within our groupings (some very large; centred on ethnicity, religion, place) we are cooperative. And it is that 'cooperative-competitive' ethos which drives the institutional mechanisms (rule of law, intergenerational responsibility, moral integrity). Such behavioural changes are the shape of a future socio-environmental compact.

---

**REASON THE SITUATION THIS WAY:**

Resource depletion and environmental change over the last 30 years exceeds all the resource depletion and environmental degradation that took place over the previous three centuries. The pragmatic challenge now confronting humankind involves cutting to a third the current rate of fossil fuel uptake, retention and expansion of the permanent forest cover, a gradual lowering of the number of consumer units (human beings and animals farmed for food), and denial of greed-and-growth fiscal policies. Hence the arguments 'for' and the reason 'why' we should study the effects, work out the numbers, and make the correct slower 'right brain' decisions.[30]

---

## FURTHER READING

Flannery, *The Weather Makers.*
Juniper, *What Has Nature Ever Done For Us?*
Collin, *Encyclopedia of Sustainability.*

# 3

# CONSUMER CULTURE:
## An Unsustainable Juggernaut

---

**Anthropocentric Exceptionalism | Population Expansion | Waste Absoorbtion Limits | Fossil Carbon Consumption | International Panel on Climate Change 2013 | Climate Change Denial**

---

The realities shaping the future include greenhouse gas emissions, ozone depletion, nitrogen concentrations and more consumption and discard by more consumers.

### THE GLOBAL BELL JAR

Artificial bell jar projects have required the participants to forego most consumer comforts and convenience; and they have never gone on long enough to include human reproduction.

In practical fact any bell jar experiment on earth represents a bell jar within a bell jar; the most studied example being our planetary orb in space, domed-over by its stratosphere, with the sun providing daily doses of heat and light.

Think of planet earth as the only reality space we can occupy, for which there is no alternative, and from which there is no escape.

All this leads to further climate warming and a gradual rise in sea level. Aspects of these changes have been wrangled earlier; here they are identified as the product of a consumer culture.[31]

Starting with the consequence of increased greenhouse gas emissions (carbon dioxide, methane, nitrous oxide). The rate of these emissions correlates with four other vectors: the mass of human presence, the upwelling of mass capital, mass consumption and mass discard. Most of the emissions — hydro fluorocarbons, carbon dioxide, toxic ozone, methane, nitrous oxide — are invisible and odourless yet lethal to biospheric stability. Smog, haze and smoke are of course more tangible as well as life threatening.

## PLANETARY BOUNDARIES

Carbon dioxide emissions from anthropocentric fossil carbon consumption is the climate change alert sounded most frequently by the people of science.

In 2009 Johan Rockstrom and 27 colleagues collaborated as a Planetary Boundaries Group to identify a total of nine interconnected life support 'boundaries' which threaten human survival: **Carbon Dioxide Concentration, Biodiversity Loss, Nitrogen-Phosphorous Concentrations, Freshwater Shortages, Ocean Water Acidity, Chemical Pollution, Land Use Changes, Ozone Depletion, Aerosol Loading**. ['*A Safe Operating Space for Humanity*', Nature 461 2009]

Planetary Boundaries is an advance on earlier work about 'limits', 'connectivity' and 'resilience' put out by the Club of Rome. However, the Global Boundary authors' inattention to 'fiscal' and 'demographic' metrics is of serious concern.

Our global bell jar is awash with an inflationary excess of resource-moving money and resource-consuming people. These are the prime factors behind adverse global change. They rank pole position in my catalogue of planetary limits. More forcibly put; a Planetary Boundaries list, which does not identify the 'boundary limits' of inflationary print-money supply to an expanding consumer mass, is seriously flawed.

The 'greenhouse gas' phrasing identifies freed $CO_2$ as the principle global warming agent. We now know from ice core studies that planetary heat retention and climate was moderated, historically, in balance with volcanoes venting and the rotting of biomass. This is now exacerbated by industrial smokestack discharges, tailpipe emissions, consumer waste discard, and the denial of $CO_2$ absorption from clear-felled, especially tropical, forests. For guidance we look to the people of science, notably those who study the proportion of carbon gases as 'parts per million' in the atmosphere.

None of my associates has a 'ppm metre'. If they did, then maybe on a good day those in the Southern Hemisphere might read an acceptable 350 ppm, whereas in the Northern Hemisphere on a bad day that could be an unacceptable 450 ppm and rising (averaging 400 ppm in mid-2013).

Carbon gas 120,000 years ago levelled out at 450 ppm. The planet was then a mere 4°C warmer overall than it is now, had very little ice, and the level of the oceans was seven metres above what it is today! This is a strong argument for cutting carbon emissions, by a large amount and promptly. Global warming, simply put, is the consequence that emanates from combining the biosphere's oxygen too quickly with the fossil carbon stock laid down over previous millennia.[32]

Of course, something we cannot decode from ancient data, is the time lag between current 391 ppm readings (up 40% on 1750) and the arrival of high tide in Central Park and Piccadilly; yet the correlation is irrefutable and irreducible. Indeed, carbon dioxide concentrations could go hugely beyond the levels reached during the last ice-cap meltdowns, to a likely 600 ppm.[33]

Extrapolations establish that a minimum 80 cm rise in sea level during the rest of the 21st century is now factual and unavoidable. What many of us appreciate at home and observe day-to-day, are the temperature hikes, droughts, deluges, volcanic and wind violence affecting our current climate; the 'unprecendended changes' confirmed as 'extremely likely' by the 2013 International Panel on Climate Change. We now have good science, we know what happened 'last time' (see the endnote) and we are aware of the potential wreckage that will be caused by greenhouse gas emissions (carbon dioxide primarily) as a consequence of a massive rise in sea level following the Antarctic and Greenland ice field meltdowns.[34]

Consideration moves to an examination of other climate change consequences. Particulates in the atmosphere insulate the ground-level biosphere against the full extent of the carbon gas aftermath, the 'greenhouse' effect. Despite this the IPCC calibrates an anticipated further likely rise of 2 to 2.4°C over the rest of this century, reaching 4°C above the 2000 level. The IPCC goes on to estimate a reduction of food production by 20%, a sea level rise of up to a metre, doubled global desertification and a hugely reduced level of biodiversity. All these causal-links fulfil Lovelock's prophesy in the *Revenge of Gaia*.[35]

No cosmic motive or spiritual force is at work. Our life support mechanism is creaking from an overuse of mineral fuels and synthetic nitrates. Mark Lynas, the author of *Six Degrees*, makes clear that every degree rise in global temperature induces significant change; and that any increase beyond +2°C (on 2010 levels) will lead to irreversible changes to the biosphere.[36]

---

## Appendix B — Oceanic Warning I
## Kiribati: Climate Change and Sea Level Rise

---

Ecological balance cannot be maintained alongside an increase in population and wealth. Excess population and stimulus capital affect the health and diversity of the biosphere and degrade the overall quality of our habitat. We know that the biosphere ticked over healthily before the leveraging effect of mass wealth was factored-in by mass human presence; in effect, before 'econometric' money geared up huge stores of conjured wealth, the exploitation of resources, increases in population and vast discards of waste. In these terms population excess and monetary leverage are situations of overload leading us into la-la land, the planet where reality is suspended. Eight to ten billion people overrun by successive hyperinflation leading to waste discard excess will strike a new biospheric balance — unwholesome and unfulfilling.[37]

The rate of human increase, technological advance and fiscal leverage has turned an irreplaceable asset, stored fossil carbons, into a climate-altering problem. As witted beings it has to be supposed that because we have got ourselves into this messy situation, we could surely find a way out — essentially by getting ourselves 'back' to a sustainably balanced future — fewer of us, lesser fossil energy uptake, and a more localised lifestyle.[38]

The current World population of around seven billion (with the most recent billion added over the last eleven years) is proportioned one-fifth thriving to four-fifths poor. Extrapolated, ten billion in the future would be proportionally as poor. Two billion during my childhood included 'only' one billion poor, within an overall situation that was environmentally damaged yet equable.

Nature, of course, does not 'care' if global population increases to ten billion; the only conscious 'caring' being that which we humans contrive; the so-called 'balance with nature' being whatever the biophysical mix happens to be at any given time. The challenge is to engage our powers of prevention and adaptation to fashion healthy, happy, smaller populations with proportionally fewer in poverty.

The human habitat is a product of two 'systems' very much at odds. Initially there was the fully integrated 'natural system' (*seasonal, cyclical, predictable*) that served humankind well up until the beginning of the 20th century. This was intruded upon by a stand alone 'fiscal system' (*variable, erratic, unreliable*) extrapolated by growing human numbers in recent times, gearing-up the exploitation of physical resources and waste discard.

Striving for sound economic 'health' attributes to the global economy a 'soul' and a 'being' with pulse and heart sensitivities, and hot-cold susceptibilities. In fact the money supply system is best regarded as a mechanical construct with no moral core. Yet — staying with the anthropomorphic inference — the market recently (2008) did 'catch a nasty cold', again, and this will reoccur with increasingly serious social consequences. Something-out-of-nothing fiscal fabrications are destined to always crumple, having sown the seed of their own inflationary chaos and decline. Despite individual, household and community attempts to 'live simply' no significant environmental improvement has been achieved.

---

A disappointing aspect of the majority of the 1970's writing (reinforced in the 1990s by the 'green growth-capitalists') is reliance on wealth generation and green-growth as the means to spend and buy a way out of the pending climatic disaster.

By 2008 the kind of economic boom those authors had in mind for 'purchasing' an improvement to the environment, had emerged; only to feed consumption, population increase, expanded inequality and climatic change!

---

All along, and available, was the knowledge and inspiration of the new-science and new-economy luminaries, writing on the subject around the 1970s: Paul Ehrlich (1968), Ernst Schumacher (1974), Rachel Carson (earlier in 1962), Lester Brown (1974), James Lovelock (1976), Fritz Capra (1975), Donella and Dennis Meadows (1972), Edward Goldsmith (1972) and Amory Lovins (1977).[39]

In different ways these people foretold what the IPCCs 2014 Assessment Report 'unequivocably' endorses, that the environmental crisis is now 'extremely likely'. It is not a matter of 'if' or 'when' but 'how much'. Their insights, based on evidence, foretells the logic of searching out and fashioning new values and a new consciousness, anchoring why behavioural adjustments and technological adaptations are essential for a wholesome continuation of the human venture.

The prescriptive 'Action and Adaptation' chapters in Part B yield fit-for-purpose social parameters, fit-for-purpose fiscal structures, fit-for-purpose energy capture from natural sources, and the engagement of eco-compatible technologies.

Shifts in the harnessing of physical energy have occurred previously; wood-fire to coal-fire, coal-fire to oil. The next shift has to be away from mainly oil, coal, gas and shale, toward increasingly more solar, wind, hydro and geothermal. This will lead to cost increases for fossil energy; in turn notably shrinking local and regional trade, and inducing a drastic reduction in the swap-goods (biscuits for biscuits) and swap-people (tourists for tourists) international trades. Families, necessarily smaller families, can harmonise their future through local food and close-to-home work and school options and natural-energy capture, with an upsurge in the social virtues of co-dependency.[40]

### Appendix B — Oceanic Warning II
### Solomons: Cargo Cult Delusion

The ability of governments to correct recent (post 1970) environmental damage has proved, thus far, to be beyond their capacity, capability and enthusiasm; a failure fed by fear of a backlash when politicians are obliged to caution electorates to not expect everlasting growth and consumption.

A prolonged faith in growth-succeeding-growth has encouraged us to put our personal consumption of energy above those of community, national and global interests. Put the matter this way: when asked to decide between a warm holiday frolic in a distant land next winter, or safeguarding the future for children yet to be born, most of us would give the matter some thought, then take the holiday! Global climatic harm inheres in the sum of our individual consumer-led preferences.

A major challenge is to attract people of wealth to look into the kidult consumerism at the centre of their ethical being; to 'find a way back' to a reality-grounded economic system; and to make judgments and arrive at commitments related to sustaining a healthy habitat largely reliant on a renewable energy mix. This challenge views continuing growth, along with the oxymoron 'sustainable development', as heedless and amoral.

Chapter 2 summarised the denial position for those of established wellbeing, the people who pass adverse reality on the other side: perceiving their lifestyle 'freedoms' in a blinkered way, shoring-up capital fortunes, and contemplating history in terms of their personal lifespan. Few who consider themselves comfortably well off would deny this.

On the evidence before them, looking back on the lives of their parents, thoughtful people with access to plenty must now concede that their generation has been over-mobilised, over-nourished, surfeit-housed and gadget-infested: and further concede that this cannot continue.

---

### Appendix B — Oceanic Warning III
### Samoa: Rich & Poor

---

Within all nations the major location shift throughout the last century has been off farms and into urban settlements; out of the landscapes of food and fibre production in alliance with nature into urban landscapes of construction, consumption and waste discard at odds with nature. Also, in the OECD context, between urban and rural there has emerged an ex-urban encroachment, broadacre urban-focused living; a high mobility 'lifestyle' located mostly on once productive farmland. Co-related outgrowths, suburban and ex-urban, emerged in sequence and expanded rapidly, along with a demand for comfort, exotic food, lifestyle accessories, and personal mobility.[41]

The crossover, with oil demand rising yet production falling, marks a move from satiation to reality-denial to situation-panic. It is a circumstance bound to inject austerity into auto-only suburbia and ex-urbia. The presumption of ever-continuing growth and continued access to mineral carbons will decline inevitably. One consequence, already, has been a minor alleviation of consumer excess in some nations of wealth (recycling, insulation, walking-buses, solar energy collection among others). But these enthusiasms, essentially pinpricks, have not produced anything like an appreciable ratcheting-back of ecologically subversive rates of carbon consumption and $CO_2$ discharge, establishing (2013) IPCC Reort "...the period 1983–2012 (as the) likely warmest period of the last 1,400 years."

The denial lobbyists see consumer control as an assault on their already noted 'freedom' to exploit energy resources and dump waste. Difficult to fathom are those well-educated contrarians and sceptics who continue to peddle misinformation. These junk-science lobbyists (*Merchants of Doubt* to Oreskes and Conway, protecting interests and avoiding disclosure) are patronising at best, monsters at worst.

It is important to wake up to the fact that tobacco firms, pharmaceutical companies, sweet-drink manufacturers and oil companies reward from their slush funds outbursts of misinformation from academics, political sympathisers and prominent personalities, when such people say something inherently profitable to their corporation. My advice: when faced with an outpouring of confusion from a denier lobbyist (contra-science contra-sustainable contra-humanity) is to hammer away with these two suspicions: 'Who rewards you for saying this?' 'Who funds your research?' — along with an improbable supplementary 'What's your IQ?' because we now know, historically, that really clever people are prone to make really stupid (socially damaging) decisions for self-advancement, laying blame elsewhere when they are proved to be mendacious.[42]

The growth-on-growth era through which humankind has 'progressed' over the last two hundred years is ending with a sequence of wake-up calls from the IPCC (established originally, an irony this, with US support from progressive Republicans). The initial concern was for the 'probable deterioration' of the environment. There followed five more reports now identifying environmental shifts of global relevance and significance as 'extremely likely'; *viz* inducing climate changes which will be damaging, lasting and irreversible. What the 2013 IPCC and the 2012 BEST findings establish, in the drive for an all-nation's consensus on climate change, has now been widely acknowledged and accepted.[43]

Historically the utopians — searching out settlement perfection — have as equally overshot perfection as the deniers have undershot.

Fortunately for humankind they, the utopians, motivate and direct our imaginations toward sustainable principles. Balance and co-operation permeates utopian thought. It is not the utopian way to foreshorten the future; indeed utopians posit that the survival of communities depends on an ability to live off nature's interest while eking out the planet's resource capital. The utopian quest is to strategise an endgame; to play nature for a satisfying harmony; to invoke self-reliance and waste-avoidance; to pursue a happy wholesome and lasting lifestyle.

Perverse to the utopian theme is the way that 'deniers' (the contrarians) sidestep science to hawk optimism and ideological hunch, the major delusion and hoodwinking of our time. Of course, over geological time, there will ultimately be an end; the sun will eventually blow-up, suck-in or fade-out, or the planet will take another king hit from a massive meteor. But the good-news prospect, for now and foreseeably, is that were there a considerable reduction of earthbound resource uptake, the end-to-end-all would be millennia away.

The bad-news prospect is that with human agencies continuing an unbridled mining of energy-dense carbon resources, the current pattern of resource consumption constitutes lifestyle roulette, with very few winners (survival of the asset dominant fittest) and hordes of losers.

---

**A challenge to put out is 'Why continue to jeopardise an attainable and wholesome human future?' And to couple this to the challenge first raised in the Introduction 'What should we now interpret as future progress?'.**

---

This reasoning repudiates Fukuyama's prognosis for the 'end of history', the end of hope, change and imagination. It embarks on a rewarding continuity; pursuing through social awareness a social democracy that sets out to pass on a livable habitat; this generation's prevention and adaptation for the future wellbeing of the next generation, ensuring that there is *not* an end to human history![44]

## FURTHER READING

Hansen, *Storms of My Grandchildren.*
Rockstrom et al. *A safe operating space for humanity.*

**PART A:** LIMITS OF RESILIENCE

Carson, *Silent Spring.*
Meadows' & Randers, *Limits.*
Lynas, *Six Degrees.*
Oreskes & Conway, *Merchants of Doubt.*

# 4

# ENVIRONMENTAL AWARENESS:
## The Politics Of Greenwash

---

Human Consciousness | Truth Evasion | Politics and Power 'Right' and 'Left' | Environmentalism | Co-Operatives | Footprinting

---

Information, awareness, understanding, this is an assimilable trinity. And from a linking together and feedback between the three the environmental conservancy of a community, a region and the planet can be advanced. Education and training for sustainable adaptation is vital.

Human continuity is a baton passed on. This take on our mortality means nothing cognitive for non-human life forms.

## ON CONSCIENCE

Other mammals, our genetic cousins bar a gnat's whisker, exist without thoughtful consciousness.

My friend Peter and I 'own' three male dogs between us – Dulux, Trouble and Womp. In a conversational moment, dogs gazing at us with apparent adoration, I asked Peter what he thought was going through their minds? "Food" Peter replied, "...and if they weren't neutered, sex."

What is poignant about his comment was my anthropomorphic presumption about the dogs' adoration of us; when in fact each animal's attachment only has meaning in terms of getting by in this, their immediate life.

Contrast that with the singular human emphasis on conscious human concern for the future of our children and grandchildren. I allude to the human ability for backward-thinking and forward-thinking, two generations in either direction, as a compelling motivation.

The philosopher John Gray (2009) puts it like this "Other animals do not need a purpose in life ... the human animal cannot do without one". Many species, ill adapted to survive in the face of change, have gone; but in and of themselves they made no intentional mistakes. Humans make cognitive mistakes all the time. 'Perpetuality' the ability to appreciate the wonderment, not only of human life but all life past, present and future, defines our being.

Management systems have transformed family enterprises into conglomerates with greater corporate control over us than we have over them, due in large part to their secretive self-serving ways.

Information leads to understanding; and beyond understanding, regulation. These are the oft-presumed pathways to control the use and overuse of climate-altering resources — but they're not working!

---

**Appendix B — Oceanic Warning IV**
**Bougainville: Corporate Greed**

---

Prior to Darwin's socio-biological reasoning, our presence on Earth was mostly based on simple belief and mysticism, customarily to maintain social control through prescriptions for moral behaviour. Important as belief and spirituality is for human groupings, in terms of practical understanding this has now been largely set aside by science and secularism.

## ABOUT POLITICS AND PLANNING

Every planning decision for change is a political decision. The more stand-alone a planning decision the more political its profile.

Unusual for a climate-change polemic, this book comes from a planner; a working life steeped in orthodox regional and local planning in developing nations.

One other thing politicians and planners have in common is that both are guided by briefings. The politician approving an education programme or a utility installation is not expected to be a qualified educationalist or engineer. Education and utilities are 'black box' matters on which they are briefed. Likewise the planner positioning schools or utility installations gets briefed to do so without being qualified in education or engineering.

This point is made in anticipation of charges being levelled about not being up to speed with the cognate sciences – who is? The reasoning offered here being that an objective approach and synthesis (the synoptic 'planning' format) is valid, viable and proven for any scale.

One thing is clear and certain; extrapolating 20th century energy-fired consumer culture and 'radiative forcing' deep into the 21st century, and coupling that to an even greater mass of population, will be catastrophic for humankind.

Knowing where matters environmental and demographic stand, as we now do on account of IPCC (climatic) and UNESCO (demographic) reportage, is basic to calibrating the production and consumption controls required to return to equilibrium. We are learning about the need to act for intervention: but nowhere near as fast as we learned previously to grow, consume, exploit and discard. The results from past mistakes are coming in; and the prognosis indicates a probable incapacity to recover from the current trend of change.[45]

Truth evasion and dumbing down continues. Dominant is the ad-machinery, promoting consumption as the human life-purpose, with a lack of truthful attention to the negative externalities, extolling the 'freedom' to do damage without paying for the damage done.

When it comes to environmental degradation truth-avoidance inhabits the headspace of the wealthy, a mendacity which separates them-us from the rest of the population. Notable are those aloof from adverse environmental change through an adherence to market fundamentalism and growth; holding to a simplistic conviction that all is well, so carry on.

---

David Fahrenthold's findings (2010) establish that US belief in climate change has fallen to 72% [50% in 2012]; which led him to identify climate change as "designed to be ignored ... a global problem with no obvious villains and no one-step solution ... a problem for somebody else at some other time."

---

The popular media (tabloids, television, radio, pulp fiction) and now electronic communication (cellular phones and the web-internet) suggests to some that we are moving on from being part of the problem to becoming part of the solution. In all its communicative ways, the modern media is tapping into the minds of youth and young adults; drawing attention to the environmental crisis ignored by their petrol-head forebears — well skewered in Rosebraugh's film 'Greedy Lying bastards'.[46]

Much confusion still gets relayed. To the fore is the crops-into-ethanol complexity. Biofuels production would be a process of utility were digesters able to consume plants root-and-branch, and all plastic, fibre, animal waste and $CO_2$ emissions, and turn the resulting soup into biofuel, fertiliser and water. But in reality the current inefficient process — growing fuel-crops on food-crop land to produce ethanol for pleasurable tooling around in automobiles — is an example of a dubious 'greenwash' motive being led on by some pretty ridiculous science.

Flawed reasoning also attaches to the food-miles posturing. Outdoor tomatoes trucked in to wintry London from sunny Spain consumes less mineral fuel and exhales less carbon dioxide overall, than tomatoes 'manufactured' in Croydon hothouses. Of course, all food (and flowers) delivered by air is hopelessly profligate in terms of fossil fuel expenditure; a situation where awareness about the energy embedded into production and delivery can inform shoppers.[47]

The media has an easy ride with food-mile arguments against transportation from distant suppliers, putting the case for locally produced energy, food, fibres and construction material; but makes hard going of the 'total ecological footprint' explanation. There are interpretative problems. Is an ecological footprint something you individually (or collectively) consider? Is the footprint best understood when expressed in hydrocarbon quanta? Might a footprint be better understood simply as an area of land? If so how does that incorporate fish-as-food productive waterscape, and mineral carbon mining from platforms over water? What about the territory, waterscape and atmosphere used for waste absorption? And what about the freshwater harvesting footprint?[48]

## FOOTPRINTING

Most of us find it difficult to envisage the spatial extent of our 'footprinting' – about five hectares each within the OECD grouping.

What we can do is recall the lifestyle pattern of consumption during the time of our youth, and compare that with today. In those terms, over half the adult population in Britain can recall 1972; which is interesting, because 1972 has been nominated as the year, in the past, when the often proposed 80% reduction on carbon emissions for today's future, was then the lifestyle reality.

Going back further, personally: during my own 1930's childhood our family of four did much jam and fruit bottling, soap making, manual lawn mowing, vegetable gardening, house cow milking, walking to school, radio listening and reading. And yes, I found both the 1930s and the 1970s to be happy decades. The seminal reference is *Our Ecological Footprint* by Wackernagel and Rees, 1996.

At its most simple 'footprinting' is understood as the all-up territorial platform an average person uses to support their lifestyle; about five hectares (twelve acres) of supply landscape being an average bandied about for the OECD context. An average? Adults living within a family income of $40,000 and a personal five-hectare 'footprint' may be considered marginally culpable.

More reprehensible is the footprint of a multi-million director, or bonused-banker to whom consumption of all manner of mineral and biospheric resources is seen to be the point-and-purpose to their profligate being. At fifty, maybe five hundred or five thousand hectares of supply landscape per consumer, they become major drivers of climate change.[49]

Then there is the matter of application, adjustment and adaptation. How does a person establish an individual 'footprint' and then 'adjust', it being a challenge to envisage shopping-around for a smaller footprint? Even the thoughtfully concerned can find the practical specifics of their footprinting confusing to understand and address, and are thereby denied the satisfaction of proving a point through their downsizing.

---

I hold to a conviction: that in the OECD-OPEC context, individuals could each halve both their personal terrestrial footprint and energy uptake without diminishing the intrinsic quality of their lives.

---

Carbon trading, carbon taxing and the cap-and-trade combination are other instruments, calibrated mainly in Dollars and Euros; all of which is pretty meaningless in terms of environmental outcome. For most of those with capital wealth, carbon constraint by price is viewed as an inconsequential extra cost. To them taxes and surcharges are mere monetary penances. They are happy to pass the emissions and discards of their consumption on to the usual burden receptors, other poorer people and the already hurting environment. Money expresses the snapshot price of everything material; but, because it is and always has been an ever-changing and inconsistent measure with no moral core, it is not capable of meaningfully 'pricing' the future 'worth' of the natural environment to us now, or to our offspring in their future.

There exists a more easily understood relationship connecting ecological conservation with eco-labeling, sometimes abbreviated as the display of GreenTick logos. Contemplate the eco-discerning Dutch shopper visiting a grocery store to purchase a non-Netherlands product, say olive oil. Such a consumer may have found the now familiar display of 'nutrition' and 'country of origin' information on packaged food intriguing and challenging; and be well pleased to have that information because of the way it informs their selection.

They would find it additionally informative, and useful, to have auxiliary labeling which showed:

1. The monetary value of any subsidy put into production of the olive oil.

2. The tariff imposed when the olive oil package crossed into the consumer's jurisdiction.

3. The amount of fossil carbon energy embodied in a product during production processing and delivery.

Eco-labelling enables discerning shopper's to select, at the point of sale, packaged foods and consumer durables with a low Sequestered Energy Footprint (SEF).[50]

A logical extension of eco-labelling for most manufactured products — particularly bulky consumer durables — involves displaying at the point-of-sale data which also informs purchasers about the energy sequestered during manufacture, the energy efficiency when in use (now common in many jurisdictions), and the energy expenditure which will be expended on disposal (allowing for the resource worth of any recovered material). Also, in these terms, how about an energy printout for a packaged holiday, for placing the kids in out-of-neighborhood schooling, for running a prison, building a freeway, a football game or a Warren Buffet jamboree?[51]

Ecological, environmental, economic and ethical information are vital aspects of life-skill decision taking. This is not only a matter of do-goodery, nor is it part of an authoritarian crusade; it is about heightening awareness and improving understanding of, and adjusting to, the effects that arise from resource consumption and waste discard during the conduct of everyday affairs. Societies still experience consumer promotion and product falsehood.

Advertising fuels an individual's desire for goods they 'want' but do not 'need'. During the 19th century Peter Kropotkin foretold the social wisdom of local production, local marketing, and self-sufficiency; stressing also the need for systems of education which argued for natural lifestyle choices; teaching what we now contemplate as a sustainable culture. One hundred and twenty years on from Kropotkin, in accord with his message, we have a sustainable advocacy: for solar and wind power generation and heat collection, waste recycling, local food production craft skills, cycling and walking, and insulation among others.[52]

Development work in economically poor countries led me to the realisation that an important factor affecting outcome — success or failure, for growth or for the environment — is power. The political obstacle to best-fit sustainable balance could be broadly categorised as the challenge of negotiating the politics of hierarchy, influence and folly. Most easily identified is the power which goes with rank: the tribal leader in Ghana, exasperated at my youthful persistence, fixing me with 'I am doing it this way because here I am the Chief!' is similar to today's arrogant wielding of power by banking, corporate and media leaders.

Additional hierarchical variants involve oestrogen and testosterone drivers, born-to-lead presumptions, and the complexities of pecking order and seniority. Alongside these variants are challenges arising from religious conviction and racial identity; whatever one set of adherents is for, the other is likely to be against![53]

In recent times, the environmental factor began to profile; not precisely a matter of being 'for' or 'against', more a matter of politicians working out how much to bet each which-way on 'energy' the 'environment' and the 'economy' in combination with variations of ideology, hierarchy, race, vanity and religion.

---

### Appendix B — Oceanic Warning V
### Fiji: Discord & Chaos

---

A first objective for political contestants, even with clean-run elections, is to secure victory by every means available: legitimately, quasi-legitimately, and at times through downright fakery.

Permeating all, the focus of a newly elected government then shifts to the orchestration of ways to get re-elected 'next time'. The run-of-the-mill top priority is fielding lobbyists, consolidating the political position, and securing continuity in office; leaving the all important matter of the next generation's wellbeing to that rare entity, the 'statesperson' with personal interests second to the public good.

There are two other points of a significant political nature. Every environmental decision is a political decision; giving rise to consequences well beyond the tenure of current office holders. Then there is the call, in the context of environmental planning and power politics, to reach beyond short-term adversarial positions to bipartisan (pan-political) agreement on the policies and problem-solving adjustments required to achieve climatic stability.[54]

It is important to also understand the massive uncaring mechanistic leverage that every expansion of money supply inflicts upon the living environment. Money in exchange for oil, following demand, led to trillions of dollars heaping-up in oil endowed locations (primarily the OPEC 'petrodictator' cartel) well away from transatlantic centres of capital generation and control.

Dubious financial 'services' (most notoriously credit derivatives and securitised debt provisioning) then led to accretions of 'shadow money' lent through 'shadow banking' to 'ninja borrowers' (no income, no job, no assets — hence no ability to pay) ruining their future. Such mismatched environmental-economic dynamics are driving a realignment of economic and political power (China, India, Brazil and some other large once-poorer nations). In all this there occurred (2008–2010) some minor downward blips in oil extraction and consumption, tracked by lower oil prices; yet oil producer's, always cushioned by their cash reserves, never stopped making profits, even when prices ebbed low.

Consumer waste profligacy at the Millennium Date rate cannot prevail for another century; even less than the time-period the Easter Islanders took to compromise their living space for their offspring. Despite this masterclass in the ruination of habitat, in most jurisdictions the political policy position remains 'trust us for we know what we are doing'; a case of short-term consumer passivity abetting the longer-term climate changes which compromise the human habitat. Always the environmental preference of choice in damaging situations is shutdown, restitution and penalty; an approach that makes politicians nervous, especially when enviro-edicts lead to job losses.

Leaders prefer to laud policies such as the production of clean burning biofuels (notably crop ethanol) even though the process is inefficient and in competition with food and fibre production. The main reason nations of relative wealth endorse crop ethanol production is, of course, because this generates jobs and feeds consumption — soaking up unemployment — thereby enabling workers to carry on being dutiful consumers and loyal voters even when the carbon input-output ratio, notably for corn and wheat ethanol, is unfavourable.

Concepts like cap-and-trade (setting environmental limits and enforcing economic penalties) attract political attention. But this is really a chalk-and-cheese comparison, evaluating damage within the 'living' domain by means of a 'fiscal' construct; using systems so unlike that there is little point in trying to correct an imbalance within one, by payments from the other. In terms of environmental stabilisation the time has come to recognise speculative (and in many instances criminal) mass-wealth creation as environmentally damaging and socially divisive. In a later 'Limits of Resilience' passage on Biospheric Economics (Chapter 7) these money-conjured-from-nothing contrivances are identified for attention.

A complication arises in that we are, most of us, implicated speculators. For example Fractional Banking Rules has allowed prospective property owners in the wealthier nations, as mortgage seekers, a loan-to-income ratio of more than ten times annual household income. This arrangement eventually ballooned out in the US to 100% 'no questions on income asked' mortgages. Along with the recent trade in derivatives this became one of the most widespread packages of uncovered debt ever devised.[55]

Adding to the political complexity. In the early 1990s Peter Mandelson spokesperson for the then marginally centre-left British Labour Party was admired for proclaiming that he was "supremely relaxed about people becoming filthy rich" under Labour (roughly the US Democratic and European Social Democrat equivalent). So much for 'environmental socialism'. At a stroke an all-party level playing field, convenient to the 'right' was set up by the 'left', with neither of the main political opponents having to lock horns over the impact of wealth on the environment. Since that time there has been a welcome shift in the perception and appreciation of environmental responsibilities, with politicians from both ends of the democratic spectrum now listening, looking and learning afresh.

Yet from both the 'right' and 'left' the majority seek to rescue the environment by increasing the cash flow into 'sustainably' slanted businesses; and to make money out of 'going green'. This is not helpful when such enterprises remain fixated on production and profit using fossil carbons for energy and the manufacture of derivatives, further straining the planet's resource-provisioning and waste-absorptive capacities. What is required is an even greater political convergence of understanding about environmental adaptation; both 'left' and 'right' policy maker's seeking the same environmental goals.

Something else has happened; the fiscal gaze which came into Transatlantic focus during the 20th century has gravitated toward the new China-Japan, Indian-Subcontinent, Singapore and Saudi epicenters. Forget Billions; think Trillions. Forget investment return; think offshore investment in resource security.

The recurring problems bedevilling political engagement in environmental affairs are that the tenure of elected office is very short, and the climate change problem very gradual. This, of course, concentrates the political mind onto termly matters and termly policies; and, in the environmental context, rewards and penalties. Rewards imply a benign after-the-event acknowledgement of worthy environmental outcomes, and are thus more favoured politically than penalties and taxes. Popular is assistance with home insulation, solar heating and solar energy collection, and some subsidy support for the provision of carbon alternatives, including the tokenism of biofuels production.[56]

Taxing carbon consumption — in effect carbon use pricing — is unpopular and regarded, in its progressive form, as penalisation. Taxes are also perversely problematic in that they create revenues that can be used in environmentally ambiguous ways. A tax at the gas station might be put into roadwork that improves vehicle efficiency a little, and encourages vehicle use a great deal. Better, for example, would be a hefty on-road tax applied to larger, older and emphysemic vehicles, a light on-road tax for smaller vehicles, and a tax holiday for approved electric, hybrid and two-wheeler vehicles more effective in reducing the burning of carbon fuels and the emission of greenhouse gases.[57]

The plunder and pillage of earlier centuries generated relatively minor and local ecological damage; and always there was the New World to expand into and mess up. But with global population now hitting seven billion expanding to ten billion, and money being printed at political will, degradation of habitat beyond biospheric limits emerges as omnipresent.

An open-thought networking of policies and practices is spreading electronically, establishing an environmentally savvy political understanding. The population at large is being moved to appreciate an environmental perspective that transcends the consumer credo of 'self first' and 'greed is good'. This profiles the worth today of prolonging the continuity of a wholesome global habitat tomorrow; in a sense ensuring our immortality, adopting and enforcing a social covenant for a healthily conserved environment for our children's children; the theme of Niki Harre's *Psychology For a Better World*.

Laudable but less assuring are 'jolly hockey sticks' exhortations for individuals to 'get with the game', 'do your bit' and 'enlist with the team'. The real world community-with-community challenge is to diversify to a deep-time reality-induced adaptation; averting the environmental hurt arising from over-wealth, over-population, and over-consumption; and setting out to ensure the livability of the biosphere we pass on to the next generation. Considered objectively, it becomes clear that it is statesmanlike political direction at the national level that must be energised to get behind the 'sustainable actions' and 'sustainable adaptations' outlined in Part B.[58]

## FURTHER READING

Dilworth, *Too Smart for our Own Good*.
Giddens, *Politics of Climate Change*.
Gray, *The Silence of Animals*.

# 5

# PEOPLE AND THE ENVIRONMENT

Population Pyramids | Neo-Malthusianism | Immortality | Manipulated
Population Change | Human Consumer Units | Urban Drift

Nothing exceeds the joy and wonder of freely desired and lovingly nurtured life. In new life, the human community celebrates and inheres.

During the low-tech phase of human development, up to say 1800, global population grew to about a billion. At that time the niche for Homo sapiens within the biosphere was assured. There then followed a six-fold increase of population during the 19th and 20th Centuries, coupled to an exponential increase in wealth and technology, bundling-up as today's climate change complexity.[59]

## POPULATION EXCESS

The seven billion mass of population (with only one billion relatively very well off) now strains the planet's carrying capacity and endangers the global habitat; notably so if the proportion of those that are wealthy increases and continues with their rampant consumption. Every take on the necessary responses – in the main stabilised money and declining population – is challenging.

Proposing, suggesting, or even hinting at reductions to the numbers living, and being censorious of those excessively wealthy and long-lived, opens entrenched debate on factors fundamental to consumerist, religious, ethnic and racial loyalties and values. These matters beg attention on a turning point of fact: namely that the consumer-discards of human origin have grown to exceed the waste absorption capability of our planet.

What is proving unsatisfactory is the 'trade', 'offset', 'tax', 'buy-out', 'penalise' and 'reward' certitude of those governments and corporations which contend that carbon gas trades and extracting cash penalties from polluters is a workable way around the debased and damaged environmental situation into which consumer humankind has boxed itself.[60]

The growth-on-growth impossibility also highlights the need for a widely effective slow-and-sensitive demographic adaptation; fewer more cherished children leading to a lesser population and lesser consumption.

Archaic beliefs, dysfunctional political systems and expanding bottom-up affluence contributes to population oversize, mostly in poor nations; generating, in effect, a partial population ponzi. The longer-living, wealth-garnering, resource-consuming and expanding human presence overwhelms both the resource supply capability and waste absorptive capacity of the biosphere.

---

**It is apparent that for humankind as a whole, achieving sustainable long-term environmental balance requires a sensitive lowering of consumer numbers over a carefully judged timeline.**

---

There are some who attempt to deflect this concern for overpopulation as neo-Malthusian; but they should take into account that in the late 18th century Malthus had only a hazy understanding of the geographical vastness and productiveness of the New World, little forewarning of the coming technical improvements in agriculture, and, of course, the mere one billion population of those times did not then live and consume into old age.

We now know about the contemporary 'limits to growth'; that the food and fibre-producing platform is fully occupied; and that waste discard is broaching absorptive capacity. The resource supply capability and the waste absorptive capacity of our 'global bell jar' has limits, which in turn delimits the number of human beings capable of enjoying a worthy-to-them quality of life. Bringing children into the World and nurturing them has always been a dominant social driver. Biospheric limits now indicate an adjustment; love for the child-singular in a technologically sustainable and fiscally steady state future.[61]

The management of population size (birth rate limitation bringing about an age pyramid slimming) as part of the quest for an environmentally sustainable outcome is challenging.

At various dates throughout the 1960–70s I worked in Sri Lanka, forming friendships through working associations with officials and their families. It was a time of relative calm. An effective malaria eradication campaign was linked into a broad literacy programme resulting in smaller, healthier and better-educated families. Prior to this, larger families were the norm; established to ensure that care and support was available in parental old age. Politically led educational and health policies worked to the Sri Lankan demographic and environmental advantage at that time.

A more recent example of manipulated population change comes from Iran. The last Shah's regime achieved a downward shift of population growth — fewer births — from an all nations' high, to a modest replacement level. Then followed the Khomeini propaganda reversal, with births again exceeding deaths by 4% through the mid-1980s.

## ON OVERCROWDING

During a time of relationship tension my then young daughters went on holiday with their mother, leaving in my care five recently acquired gerbils. One escaped and bit into an electric cable, so there were then four.

One morning a week later the cage held several more, maybe eight; the girls will be excited! As I was about to set off on a lengthy field trip the gerbil's cage was put in the garden shed with a drinking bottle and a four kilo bag of food to await the return of the children. Twelve weeks later – the girls had forgotten them – the food was all gone, and in the confined cage were dozens and dozens of miserable gerbils.

The point being made is simply that lives of mere existence in mass, such as ten billion people would have to endure within our planetary 'cage' would not be dignified.

Also worth noting, is that some South Eastern Asian nations have voluntarily constrained fertility rates — Japan, Hong Kong, South Korea, Singapore — the main downside being that they are all now obliged to manage a large and looming pensioner-age cohort.

None of these reductions in fertility has induced an outpouring of sorrow for those not brought into this world. Nor indeed does the current lower birth rate in Russia, China and throughout Europe, produce significant observable lamentations of social denial. At base, every additional human consumer born into an already crowded planetary space, impacts adversely on both finite-irreplaceable resources and the living-renewable resources available to support human existence overall.[62]

There is an applied concept in pastoral farming, the 'stock unit', which is worthy of technical consideration as a means to evaluate the capacity for human 'consumer units' in any terrestrial context. It also offers a down-to-earth way to further our understanding about ecological footprinting. For example, the human consumer unit situation for desert-like N'Djamena in Chad can be compared with also desert-like Phoenix in Arizona.

## ANTHROPOCENTRIC NITROGEN

To the authors of 'Planetary Boundaries' (aside in Chapter 3) interference with the nitrogen and phosphorous cycles is as serious as atmospheric $CO_2$ and methane accumulations on account of the nutrient over-enrichment of the waterways and oceanic ecosystem (the hydrosphere).

Nitrogen is 'fixed' from its atmospheric source by the Haber-Bosch process which manufactures reactive nitrate fertiliser and nitrites. The fertiliser runs off in solution polluting rivers, lakes and the oceans at about five times the limit recommended in 'Planetary Boundaries' "...erode(ing) the resilience of important Earth subsystems ... (which) thus directly increases radiative (greenhouse gas) forcing."

On analysis this predicates altered water and nitrate resource uptake regimes; respectively up for Chad and down for Arizona. Not only is the consumption of water and nitrate resource uptake for the average life lived in each place different, this comparison can also factor in that average life expectancy in Arizona is twice the Chadian.

High-end consumers have in the past survived short-term change in ways the wealthy population of Phoenix substantially drew down the Arizonan aquifer, then looked to import water from elsewhere. Low-end consumers like those in Chad are hit harder day-to-day yet survive over the longer term; coping with fluctuating water availability.

This leads to the conclusion that an across-the-board proportional reduction of all populations is not advocated. What is called for is population balance — settling upon a population and consumption stasis that can be sustained.[63]

If it could be achieved, downsizing the rates of carbon and nitrate consumption by say 50%, would emphatically reduce individual ecological footprinting. A valid and achievable objective would be to lower to a third the twelve acres per person for high-end users, to around four acres per person. However, in terms of official policies and practices, neither the carbon reducing nor the footprint-lessening concept has been brought into effective play.

In the context of this review of 'People and the Environment', lowering the density of human and domestic-animal land occupation throughout all jurisdictions would achieve human continuance and wholesome survival over the longer term.

As noted earlier, for the first time in human history, over half the global population is urban, moving in greater numbers to mega cities. One consequence of rapid urban growth, and a looming humanitarian tragedy, is that the increased proportion of those in poverty are urban slum dwellers distant from food and fibre sources, taking from rather than living within the open land and water territories.[64]

Imbalance arises also in relation to the loss of biodiversity, particularly the feral animal and wilderness plant communities. They have no conscious self-voice, yet clearly in terms of quality-of-life, it would be better for them if there were fewer of us. A growing human population further increases the extent of overall human activity and the number in poverty. Within a reducing human population the poverty median trends toward proportional improvement, withal improving the survival of other co-dependent species.

---

### Appendix B — Oceanic Warning VI
### Tarawa: Population Excess

---

After world population reached four billion the demographic situation, relative to the biosphere, started to grow into an imbalance. Add to the mix unprecedented capital expansion, limitless carbon energy and nitrate fertiliser supply, technological impetus, and two world wars; and of course the environmental-human mismatch increased. The human majority continues to be ill informed about our reliance on the thin-coated, solar-warmed biosphere for our sustenance.

Planetary absorptive capacity, relative to discards from the global human population, is bound by absolute limitations of biospheric resilience. The current seven billion consumers (one-fifth high end) constrict and degrade biospheric wellbeing through their exacerbation of climate change.

This is especially the situation while there is an expanding uptake of mineral carbons and fixing of nitrogen, all of which contributes to the greenhouse gas accumulation which leads to environmental degradation. Ten billion plus population would simply constrict the average quality of human existence; probably for twice as many people twice as severely.

It is worth noting, from recent progressive examples of birth limitation, that both official and unofficially endorsed population lessening not only constrains population growth, there is also an overall quality-of-life improvement. The population limitation issue (primarily a matter of sensitively managed soft-style fertility limitation) is examined in *Part B: 'Actions & Adaptations'* — Chapter 11 'Economics and Demographics'.

---

**Repeating:**

**Absolutely nothing exceeds the joy and wonder we experience from nurturing one or two freely desired and thoughtfully planned progeny.**

---

## FURTHER READING

Wilson, *The Future of Life*.
Lynas, *God Species*.
Lovelock, *The Revenge of Gaia*.
Erlich & Holdren, *Population Resources Environment*.

The nuances of demographic transition are ethically and politically complex, the responses to fertility interventions slow, the economic implications challenging. For a lucid overview consult Part 3 of Jeffrey Sachs, *Common Wealth* (2008).

# 6

# TECHNOLOGY AND THE ENVIRONMENT

Embezzling the Future | Techno-Negatives and Techno-Positives | The Biofuels Delusion | The Challenge of Downsizing | Fourth Generation Nuclear | Urban Concentration

Economic, ethical and socio-political factuals criss-cross this chapter. In order to rationalise that complexity the current focus is on identifying the techno-negatives and techno-positives, which impact on the environment. This is where hindsight gives way to some proactive foresight, followed-up in *Part B: 'Actions & Adaptations'* (mainly Chapters 12, 13 and 14) as practical adjustments for stabilising climate change.

Fiscal expansion and the growth of consumer numbers has geared-up carbon extraction and nitrate manufacture; and this has knocked-on as excessive atmospheric carbon gas and hydrospheric nitrates, global temperature increase, and sea level rise. These are all fired by a fiscally supercharged, consumer-driven and oversized human mass.

Technological advance is fiscally leveraged, chasing an illusive human dream. This expansion of consumer numbers and fiscal profligacy is reflected in the rate of fossil energy use and adverse change to the biosphere; with the poor and the least culpable most affected by climatic spikes, sea level rise, crop failures, reduced biodiversity and overcrowding.

Except for the still stone-age poor, people in most nations (emphatically the case throughout the OECD and OPEC) currently enjoy an unprece-dented technological leavening — their gizmos, food selections, mobility, comfort and lifestyle accessories. In the main this is carbon fuelled simply because oil gas and coal are a reliable, convenient and affordable 'dense' form of stored energy ready to hand.

In my small community the excesses I am prepared to admit to are marginally improved upon by a virtuous few, and undershot by a conspicuous majority. Excesses for correction arise, to a lesser extent, in Third World situations. The overall conclusion is that although there are Third World resource use and discard failings and population overgrowth, the carbon gas damage leading to climate change for poorer populations pales to insignificance alongside richer-nation and richer- people resource overuse and toxic discard.[65]

More technologically serious in poorer nations are large-scale accidental and sometimes deliberate waste discard activities (Union Carbide chemical-spill disaster in India), mass toxic dumping (Trafigura toxic chemical slops shipped to the Ivory Coast), and the probable offshore scuttling of nuclear waste and toxic sludge containers.

## PURLOINING THE FUTURE

In the past I admit to and identify with the embezzlement of my grandchildren's future in ways I now strive to reform.

We all, to varying degrees, display Consumer Behaviours detailed in the **Appendix A** construct. Other indications suggest that because climate change manifests as a 'slow' disaster we are tardy about its urgency, because of the following complexities:

- **Complacency:** An 'elsewhere' matter: faraway and far-off in time.
- **Compassion fatigue:** Weary with the fight for environmental justice.
- **Denial:** Lack of acceptance that climate change is for real.
- **Why me or us?** The inequality of the climate change responsibility.[66]
- **Impotence:** There is a problem, but there's nothing I can do about it.
- **Complexity:** Not knowing what ppm and temperature variables mean.
- **Overstatement:** The 'sky's falling in' when, patently, it is not.
- **Misplaced optimism** about the buy-out and techno-fix potential.
- **Apathy:** Unwillingness to give up established lifestyle arrangements.
- **Contrarian Propaganda** overcome by psychic torpor.[67]
- **The 'normalcy bias':** Accepting what we have as normal.
- **Inattentional blindness:** Inability to see wrong in what's going on.[68]
- **Pointlessness:** Perhaps we live in a world without ultimate meaning?
- **Inevitability:** Money will be printed, populations will increase, carbons will be mined, carbon gases will effuse, temperature will rise, and...

There is little policy movement toward downsizing and better utilising the fossil carbon uptake: *vide* the continuing dominance of gasoline guzzling automobile manufacture, a ton of metal and plastic mobilised with shameful payload inefficiency.

Out of such inefficiency arises a prospect. This is based on the fact that overall only 13% of the energy derived from fossil carbon uptake is gainfully utilised; putting before us the potential to considerably improve the effectiveness of our fossil energy capture-and-use.[69]

In these terms we are confronted by the challenge of a reduction (halving the uptake) and an improvement to the utility (doubling the gain) from future fossil carbon utilisation (an advocacy furthered in *Part B: 'Actions & Adaptations'* — Chapter 13 'Goals').

Impressive technological wizardry exists, but there is little movement toward collectively downsizing fossil carbon uptake and nitrogen fixing; *vide* the continuing dominance of oil consuming automobile operations and the bizarre application of synthetic nitrogen fertilisers to ethanol crops. Recognition of the effects — ambient pollution, resource depletion, toxic accumulation — highlights the need for change. The presumption of an ever advancing consume-and-discard modernity highlights a biospheric imperative for constraining fossil carbon supply and improving the efficiency of fossil carbon use.

In the past humankind has shown guile (*slickness*) dominating nature for comfort mobility and convenience, leveraging power through asset creation and accumulation. What is now clear in the modern era is that this is a play of two halves. In the current second-half the challenge is to become wiser and more clever (*smarter*); working with and understanding the reality of human dependence on nature and making do from localised modes of life support which are less dependent on fossil carbon energy.

The forthcoming *Part B: 'Actions & Adaptations'* emphasises a slow living (Soft Pathways) approach; the principal techno-environmental components being: smaller scale projects, selective use of environmentally proven technical accessories, local food fibre and energy production, and waste avoidance. The main practical adjustment involves kicking the squander-and-forget carbon energy consuming habit into touch.[70]

Energy supply lies at the heart of every neo-modern, neo-com and techno-electronic venture. Not only does carbon energy currently fuel the cars, heat the homes, and energise the stoves and refrigerators; carbon energy is embodied into automobiles during manufacture, into the home during construction and into all consumer durables. Mineral carbons are also the main ingredient in the manufacture of paints, plastics, electronics, medicines, fertilisers and fabrics. Carbon energy is also absorbed in food-swapping (food miles), place-swapping (tourism), and in working, schooling and entertainment. Lessening the supply of earthbound carbon fuel into the consumer mix involves a transition, inevitably a slow-start then abrupt-transition toward renewables.

Overall there is a cynical under reporting of crude oil reserves. This drives another nail into what could become our biospheric coffin. And worse, if a perceived future energy shortfall is met, less efficiently, from heavily polluting tar-sands, oil-shales and brown-coal, the Millennium plus 4°C temperature change will arrive that much quicker.

A society rendered dysfunctional by not reducing its level of fossil energy dependency is a horrendous prospect.[71]

---

**The necessary denial for wealth-consumers is a mineral fuel supply which tracks toward an at least halved uptake used with doubled efficiency (Chapter 12: Resource Conservancy with Waste Conservancy); supplemented by natural energy supply systems.**

---

What is the most optimistic default position? Additional to the 'halving uptake: doubling efficiency' mantra, there is some prospect that $CO_2$ gases and heat-trapping particulates from bulk manufacturing industries could be infused underground or undersea by carbon capture and storage (CCS); keeping in mind the fact that CCS systems require large inputs of energy to power them along.[72]

This leads to a more detailed consideration of the politically popular 'cleaner' liquid fuel alternative, biofuels produced from plant feed stock. Now established — sugar cane (Brazil), corn (USA) and palm oil (Tropics) — the biofuel approach has received widespread endorsement. But, in the likes of Brazil, commanding the use of drudge labour to provide a luxury fuel generates social unrest; and in the United States the use of powered machinery to produce fuel from fertiliser-boosted corn represents technological and land use folly, notably so in growing seasons (like 2012) of widespread drought. Biofuels derived from whole-of-plant and cellulosic timber waste, using a heated catalytic, shows some promise.[73]

More harebrained, although appealing in prospect, is thermal depolymerization (TPD). Old clothes, tyres, all plastics, cellulosic waste, and raw sewage the feed stock for this fanciful mill. The major complications being:

1. The invention of a magical catylitic converter.

2. A continuing supply of feedstock, which perversly, as already noted, would be largely fossil carbon in origin.[74]

Another energy generating suggestion involves the production of liquid hydrogen by some, yet to be fashioned, efficient process. Here, a further caution: most of us have done school lab experiments where small amounts of energy has been applied to separate hydrogen and oxygen, and recall advice about the explosive danger of the hydrogen produced. Good advice, mostly observed in the breach during the school lab! Danger exists at all stages of bulk hydrogen production: separation, compression, containment, delivery and consumption. Vehicles powered from hydrogen are costly to produce, exclusionary to own, and really spectacular accident material. The future for bulk hydrogen production and uptake remains some way off.

There are also nuclear options, at least for a while; particularly the 'fail-safe' Fourth Generation (Generation III+) plants, the modula pebble-bed option, and the change from uranium to thorium oxide feedstock. The attraction of thorium installations includes abundance of feedstock supply, lower radioactivity, less waste, less adaptable for weapons-grade production; all coupled to the ability to consume stockpiled waste from uranium-fed reactors. Another major positive is that the smaller, safer modular nuclear plants of the future will be able to provide energy free of $CO_2$ emissions ('decarbonised') from facilities adjacent to factories, settlements, and fixed-line transportation.[75]

Some jurisdictions reject fission-produced power, although relative to the ubiquitous ravages of carbon dioxide the fallout from nuclear break-down has been proportionally minor. Nuclear ranks carbon-clean relative to coal and is a multi-mega producer of energy gram-for-gram; mainly as heat converted to electricity. Breeder-reactor power is more distant, politically fraught, and demanding of stable and secure government.[76]

### Appendix B — Oceanic Warning VII
### Malima: Techno-Frailty

The use of plastics as bags and wrapping is widespread in poor and rich countries. The resource input to their manufacture is mostly fossil carbon. There is a minor switch to bio-carbon plastics. Although even when fashioned from vegetable starch, energy gets used during the growing and harvesting of the feedstock and during manufacture, the saving grace being biodegradability. One advocacy is to use recyclable plastics and, of course, to recycle them; yet worthy as this may seem it has failed to dent continuing production. The substitute container alternatives (cans and glass) also consume energy during manufacture, and require additional energy to be recycled. Avoiding packaging altogether, string bag shopping, and the use of reusable packaging (glassware) are good options.[77]

There remains a practical difficulty: about a third of the plastics manufactured are vinyls bound into consumer durables — computers, vehicles, cell phones — and the vinyl sheeting used in construction. These are products alien to nature; and although relatively inert when in place, they produce unassimilable toxins when burned, and they can only be recycled at considerable monetary and energy cost. The appropriate techno-environmental attitude to plastics is to use those that are recyclable, and to take up natural alternatives.

Food production involves agriculture, pastoral farming, and fishing; processing, storage, packaging and transportation; point-of-sale presentation and consumer purchase; preparation for consumption; then off-the-plate disposal of leftover scraps and human waste disposal. The food calorie energy ingested relative to the overall industrial energy put into food production, transportation, storage, sale, preparation and waste disposal, is very low — miniscule. Expert writing is available on this matter; here I attempt an overview.[78]

The technologies engaged in food production range from machines to chemicals, the social forces range from new politics to ancient rights, and the energy input challenge includes reversion to hand labour on the farm and sailboat fishing on water. Farmers reason that if they avoid mechanisation in order to 'go green' their bank balance 'turns red'; and even though fishers extract their profit from a 'free' resource common they also reason, within the current market system, that 'hook line and sinker' sailboat fishing is not viable. Farmers and fishers alike push for bulk throughput, adding value through processing and packaging, then setting out to maximise distribution and sales.[79]

The best way to appreciate the techno-balancing involved is to compare on-the-lips food-calorie benefits with the industrial calorie costs of production and preparation, consulting the groundbreaking work of Kenneth Baxter (1989) along with Michael Pollen's recent *Omnivores Dilemma* (2007). Most dramatic is the energy cost of mechanised feral fishing, which is little more than primitive hunter-gathering at sea; powerful boats trawling miles of sluggish gear, refrigeration, delivery, cooking; all for tiny food-calorie gains.

Adjacent to feral fish-for-food supply, in terms of inefficiency, is grain-fed meat supply. Best in terms of food efficiency is locally grown, marketed and consumed grain, fruit and vegetables; contracting and converging upon an easily agreed corollary, the health benefits of less-protein, low-fat, near-vegetarian diets. This also translates into a low mineral-carbon embodiment in food production, and close-to- consumer 'patriotic' food production promoted in the modern, yet actually quite ancient, Farmer's Market movement.[80]

Most of the non-solar energy embedded in food production — all that mechanisation, climate control, pest control, processing, storage, transportation, preparation — comes from fossil fuels. Reference is made elsewhere, and here again, to the utility of point-of-sale information about subsidies tariffs and embedded energy for food and fibre product. An extension to that listing involves the provision of input-indicators for non-solar energy input; remembering that as in the case of consumer durables, carbon consumption goes on during the transportation, storage, and preparation of food products.[81]

Rounding off this foray into food technology is identification of purposeful backyard and allotment production of fruit, vegetables and eggs; inducing all manner of energy-saving fitness-induction and feel-good spin-off; along with rainwater harvesting, composting, food swapping, sharing excess, and community marketing, all in line with a 'Love Food Hate Waste' campaign.

Urban considerations tend to dominate this discussion of 'technology and the environment' because the urban realm is where most manufacturing occurs, where the bulk of that technology is applied, and from whence the mass of air-borne and water-borne waste is discarded. The complication, notably for societies of wealth, is that automobile use has abetted the compartmenting (exclusion zoning) of settlements into separate areas for working, schooling, playing, preying and sleeping.

Citizens, previously automobile-reliant in order to move between zones, are now faced with the challenge of working out ways to adapt to the mobility constraints dictated by reduced gasoline supply. Options for adjustment to the inherited pattern of 'zoned' urban accommodation are outlined in the *Part B: 'Actions & Adaptations'* (mainly through Mixed Use Development, known as MUDs, for urban areas — Chapter 13 'Goals'), and from an exploration of the personal conservancy options outlined in Chapter 14 'Targets'.[82]

The Actions & Adaptations detailed in Part B aim to cut back heavily on individual and collective footprinting; primarily by transitioning to a reduced level of carbon consumption and carbon gas emission (in the atmosphere) and by lessening the level of nitrates and carbon gases in solution (in the hydrosphere).[83]

With the application of statutory powers and protocols such reductions would aim to link-up with resource conservancy, environmental cleanup, population limitation, and climate change prevention and adaptation. Here it is relevant to observe that technical transitions take time, and human adaptation has been tardy. Slowing the annual rate of incrementally increased fossil energy uptake has never yet occurred in recorded human history. The reality is that reducing fossil energy uptake remains both an absolute imperative and, to many, an improbable objective.

## FURTHER READING

Hardin, *Living Within Limits.*
Roberts, *The End of Oil.*
Pollen, *The Omnivor's Dilemma.*
Christenson, *The Innovator's Dilemma.*
Sennett, *The Craftsman.*
Matheson, *Green Chic.*

# 7

# BIOSPHERIC ECONOMICS:
## Perpetual Growth –
## Perpetual Debt

---

The Growth-On-Growth Paradigm | 'Rational' Economic Behaviour | Fiscal
Folly | The Money-Supply Tsunami | Leverage on Resources | 'Too Big to
Fail: Too Big to Save, Too Big to Exist' | The Economic-Entropic Congruence

---

The Chapter 4 review of political understanding around climate change
examined the relationship between naturally functioning biospherics
and the humanly contrived economic system. This has been shown
to be a relationship where people of wealth may admire the complex
living environment, yet more ardently covet their cash. From the 1970s
onward the expansion of this liquidity encouraged growing populations
to attain a dysfunctional and ultimately delusional domination, with
the metaphorical oxygen of free-market growth producing a very real
effusion of climate-altering gas.

The discard effusions accumulating from this egregious synergy have
grown to the point where, were it possible to take the riches of the one-
fifth of the global population of wealth out of the equation, the climate
change problem would be solved.

Acting against the collective human interest on our finite planet are
several money-shuffling processes, which, between 1970 and 2008,
created a fiscal tsunami. This faux-wealth was conjured up from derivative
trading, interest rate fixing (the Libor rigging), hedge-fund profiteering,
shell company fabrications, currency shunting, share speculation, the

carry trade and, to an imperfectly known extent, institutionally veneered insider trading. These contrived processes fuelled consumption, in direct opposition to the principles of sustainability. In terms of environmental degradation, Zency (2013) put the situation thus "… sucking low entropy from [the] environment and excreting a high-entropy wake of degraded matter and energy". We have grown to understand all this; yet continue with a fiscally leveraged pattern of consumption.[84]

---

### Appendix B — Oceanic Warning VIII
### Nauru: Economic Collapse

---

From the dawn of the industrial revolution human beings learned to rationalise economic behaviour to advantage: one by maximising gains through the exploitation of profiting resources, two by evading responsibility for waste discard. That mantra has morphed into a psychopathic profit-taking from resource exploitation, along with a disregard for the damage-costs associated with free-to-the-environment waste disposal.[85]

## THE GLOBAL FISCAL ANCHOR

A Full Gold Standard operated in Britain until 1914, the beginning of World War I; after which it was modified into a variable Gold Bullion Standard.

Most Gold Standard attachments ceased in the 1930s as nations worldwide sought to print money to alleviate economic recession and prepare for World War II. John Maynard Keynes, sensing the need to prevent the excesses of inflation, got around (1930s) to suggesting the creation of a stable international currency unit, the Bancor, which, along with the Chicago Plan/New Deal, did not catch on. The United States continued notionally to support convertibility of the US dollar into gold at $35 an ounce up until mid-August 1971.

From a glance in today's (2014) newspaper gold is over $2200 an ounce: that's a 60 times increase! Repeated inflation over the last 40 years ('print' money to cover debt obligations and further production and consumption) feeds systemic financial failure.

Creating mega-wealth out of almost nothing (fiscal alchemy) took-off when computer usage was linked into poorly regulated bandit-banking systems. This led to the trading of credit derivatives and their securitisation in a parallel 'virtual economy'; Warren Buffett's 'financial weapons of mass destruction'. Also, over the last century population multiplied from two to almost seven billion, while fiscal capital multiplied fifty times.

---

**Clearly there is no way monetary and consumer expansion can continue indefinitely. The recourse now open to all societies is to 'get wise', 'slow down' and stabilise; reacting to the fact that money supply, climate change, sea level rise, population growth, global wealth, and indeed oil extraction and oil prices, are all in lock-step.**[86]

---

One hundred years ago the seasonal cycles were constant, with gold the medium of monetary equivalency courtesy of the Gold Standard. Those certainties are now much altered; fired along by the post-1970s expansion of fiscal liquidity and the rapid doubling and redoubling of population. Modern communities of wealth are aware of the environmental degradation that is taking place; yet uphold the illusion of being part of a never ending resource supply, a saviour fiscal system, and servant labour demographics.[87]

First, lubricated by money supply, came population change; the movement of peoples from Africa to the Americas, followed by European and then Asiatic migration throughout the rest of the World; gradual at first by sailing ship, then by steamer, and eventually by jumbo jet. Environmental change succeeded population change; with forests milled, land cleared for farming, mines dug for minerals, settlements formed. Later came the slumps and spurts of economic change, each backward slump overarched by a succeeding onward spurt.

Much more than was the case within the Industrial Era, in the Electronic Age it became possible to become wealthy using money conjured-up Weimar-style to 'invest' in business growth and consumer expansion. The mental state of the high-net-worth individuals thus created being akin to a seldom-admitted guilt, reflected in the public pitch of their philanthropy. Worthy as this is intended to be, it seldom set out directly to relieve bottom-end poverty. What it achieves is some relief of disease-misery poverty. Neither representative democracy, nor the market, has ever overcome desperate poverty.[88]

The point in noting these rates of change is that despite population expansion (the last 1 billion in just twelve years) and a manifest degradation of the environment, societies have cemented in place a belief that 'somehow' all will be well; saviour cash will be printed by Governments or conjured into being by the fractural banking system to keep turnover rolling.

All along, borrowing today for tomorrow's comfort and security has held constant through successions of population overgrowth and resource stripping, an article of faith. This has encouraged government's to assume that resource depletion, population congestion, and now climate change can be overcome by manipulating the economic and statutory levers to 'buy' a solution; with the credit granted today paid-off in the never-never or, more likely, inflated into insignificance. Financial intervention mostly does little more than slap band-aids onto a wounded ecosystem.[89]

## FISCAL INNOCENCE: *circa* 1950

Opening a savings account, my first visit to a Trading Bank, led me to work out that banks appeared to provide a useful social service along these lines.

I reasoned that people, like myself, put our cash into the bank for safekeeping, and to earn a little interest. Then, gathering up all the customer money, the bank lent it out, at a higher rate of interest of course, to borrowers who wanted to buy houses and start businesses.

Little did I know that even then banks 'lent' money they did not have in a process known as Fractional Banking – counterfeiting up to ten times their day-to-day cash-asset holdings as loans.

In August 2012 the IMF sanctioned publication of a Working Paper by Jaromir Benes and Michael Kumhof *The Chicago Plan Revisited* proposing the end of Fractional Banking, requiring banks henceforth to fully back their liabilities with state-provided money.

The first of Paul Grignon's animations 'Money as Debt' (2007) illustrates the ongoing ability of banks, now more than ever, to synthesise money. Refer also to Endnote 87

It is now apparent that the wondrous biospheric system evolved by nature, and the alluring easy-money 'freedom' engineered by humankind, amounts to a contra-synergy. After Homo sapiens evolved speech and the ability to reason, they-we morphed into consumer capitalists with our wider 'wants' always in excess of our fundamental 'needs'. Not merely storing-up food and fuel to cover seasonal fluctuations, but garnering an excess of resources and money for personal security, as indulgence, and as a means to command the servitude of others.

With the coming of the printed word (storing ideas) and printed money (storing wealth) the attractions and opportunities for hoarding and domination increased.

The worst of it is that governments of the European and Settler Society kinds are doing now, to future populations, what ordinary individuals would never do to their children; freighting each and every individual with an enormous financial load that warps the whole of 'planet ponzi' into an indebted future. As recent as 2010 this led Tony Judt to predict that "[f]or the foreseeable future we shall be deeply economically insecure."[90]

## FISCAL LIQUIDITY: 2007

**Real Money 1%:** Notes & Coins
**Budget Money 11%:** Cheque accounts, credit card debt, at-call debt.
**Derivatives 75%:** Futures, options, debt insurance, credit default swaps.
**Securitised Debt 13%:** Mortgages, bonds, packaged mortgage bundles

[From a diagrammatic construct by Brian Gaynor September 2008. For an authoritative account of the global financial crisis consult the July 2009 edition of the *Cambridge Journal of Economics*.]

## SECURITISATION OF HOME OWNERSHIP DEBT

**Mortgage Bonds:** Collections of like-kind mortgage obligations packaged for on-sale.
**Collateralised Debt Obligations:** Bundles of 'mortgage bond packages' for on-sale.
**Synthetic Collateralised Debt Obligations:** 'Insured' bundles-and-bundles of sold-on CDOs for which insurance didn't really exist as there was nobody to pay out when they started to rot.

The early 20th century saw the emergence of trans-national banking and credit advance, which abetted the expansion of both private enterprise and government projects. The liquidity to do all this was created so effortlessly that it has become a challenge to now acknowledge that there occurred a hollowing-out of resources fed by synthetic money supply; wealth accretion provoking resource plunder and waste accumulation. Government control over money supply has been subverted by fractional banking and debt-geared financial service systems.[91]

Globally, synthesised money supply and free-rider discard externalities reached a peak just before the 2008 economic meltdown. As in 1929 this followed an accumulation of top-end wealth in a situation of regulatory permissiveness, notably the failure to read signs of banks over-lending money they did not have for a housing market unable to meet its debt obligations.

## BACK IN THE 1970s

My 1973 edition of Alan Gilpin's *Dictionary of Economic Terms* contains no definitions for 'Sub-prime', 'Derivatives' or 'Securitisation'.

My 1976 edition of the Holister and Porteous *Dictionary of the Environment* contains no definition for 'Climate Change' or 'Global Warming'.

Prior to 1970, the biosphere's capacity to absorb $CO_2$ effusions from four billions of population was of no great concern. After 1970 clever econometricians found ways to conjure-up enormous quantities of fiscal liquidity. This was followed by an increased resource uptake (notably carbon), population explosion and eventual climate change; paving our descent into a self-fashioned biospheric hell in a fiscal-fuelled handcart.

Paradigmatic resource looting began in the early 1970s with the expansion of self-regulating supply-sider economics. Prior to that time capital adequacy rules (about cash-to-assets) were partially successful in establishing the amount of cover that had to be held by lending institutions relative to the loans they made.[92]

During the 1970s Derivatives, and the process of their Securitisation, were manipulated and geared along by econometricians to the point where the amounts held by the financial services industry to support the loans they were making became insignificant, while the amounts of money they did not own, but were 'lending', became gargantuan. John Lanchester (2010) gives global economic activity at the Millennium Date as $36tn, which doubled to $70tn by 2008.

Just before the bubble burst, the Derivative and Securitised heist had amassed four-fifths of global liquidity as US dollars, much of it unsecured.

The leverage which fiscal liquidity brought to bear on finite earth-bound resources, notably mineral carbons (the dense energy of convenience) has been massive; with unrelenting money supply now barrelling-on to fund both environmental disaster and a major fiscal blowout.[93]

Relentlessly, the increase of a wealthy global population has induced expansive oil mining using already paid-up investments in infrastructure, leading to excessive carbon dioxide discard, soil and species loss, and aquifer depletion and rainforest exploitation. Over the past four decades the world has moved from a merely flawed circular exchange system, to a perverted set of financial congeries.

## MONEY
## A Store of Value or Faith?

A Texan rolls up to a classy boutique hotel in a small New England town and enquires about a deluxe suite. The innkeeper is about to go out, so asks the Texan to leave him with $100 as security, hands over the key, and sets off into town.

The innkeeper calls on the butcher and gives him the $100 for meat delivered recently to the hotel. The butcher sees his farmer-supplier in the street so rushes out to pay him $100 for animals delivered to the shop. The farmer then goes into the cafe to pay the local prostitute $100 for services rendered. The prostitute then goes to the boutique hotel just as the innkeeper returns, and pays him $100 for rooms she has used with clients.

At that point the Texan strolls down the stairs and informs the proprietor that the suite doesn't suit him, is handed back his $100, and drives away.

The innkeeper, butcher, farmer and prostitute have discharged all four $100 debt obligations by simply passing around a pretty piece of paper printed on the authority of the US Treasury.

The linked together global consequence is now apparent. Unfortunately, governments failed and still fail to acknowledge the gravity of the walled-away economic imperialism, which shields banks, investment agencies and the very rich. The Chicago School (Milton Friedman), the Californian Ideology and the Washington Consensus have contributed not only to a narrow channeling of wealth to the 'top one per cent'; they have dumped the unwealthy 'bottom fifty per cent' with a debt obligation so complex it will never be understood, and so vast the debts will never be discharged![94]

We look for culprits. Aside from easily identified felons like Stewart and Madoff, and the delusional CEOs of Enron, it has been difficult to isolate the specific wrongdoers who initiated scams. Catching perpetrators, real persons, amounts to a blindfold search in a labyrinth of co-perjurer's. The recent upwelling of toxic debt (a re-run of 1929) with no viable creditors of last resort, have been largely subsumed as 'errors of judgment' and 'simple mistakes'.[95]

Culpable investment banks were quick, and rather happy, to be identified. They relished being treated as 'too big' to be allowed to go under, so 'let's bail them out', that governmental socialisation of banking mistakes using 'printed' money. All along, the specific problem was that big banks were duping individuals into borrowing conjured money; the banks receiving state funding for their lending 'mistakes' when the jobless borrowers defaulted.

In this context 'Too big to fail' should have been handled as 'Too self-serving to save'! Worse, what transpired when the big banks failed was seldom a matter of clear and certain trotters in the trough, it being difficult to bring specific mathematical manipulators to account.

## RESPONSIBILITIES AND RIGHTS

The 'responsible' economic decrease movement (decroissance) aims to contain monetary expansion to save the planet from resource depletion and emissions waste; but with no position on poverty and population overload.

The expansionist economic 'rights' movement (neo-Chartilist) would promote inflationary stimulus to money-up people as consumers. The neo-Chartilist's have no position on biospheric waste overload.

Paul Grignon's *'Money as Dept III'* animation (2011) adds to the decroissance advocacy. Refer also to the IMF 'Working Paper' *The Chicago Plan Revisited* (Benes and Kumhof).

Furthermore, in the search to place blame, it has been identified that the top players moved between an independent financial sector 'to make a killing', and their government treasury service 'for the public good', in a circuitous poacher-turned-gamekeeper exchange. Such persons are not drawn to find fault one against the other.[96]

High-octane liquidity quickly vaporises. Banks distrust banks, insurers fail their clients, retailers avoid producers, and national and local economies get sideswiped.

As a result some reduced-wealth consumers were obliged to holiday nearer to home, to forego fancy imports, to buy second-hand, and shop locally. Unfortunately, for the environment, these benefits were counter-offset by oil suppliers responding to the falloff in demand by dramatically lowering their fuel prices, thereby maintaining oil supply and a continued $CO_2$ outpouring. The core driver within the oil market is not, and never was, social good; it is, and remains, supplier profit.

---

**Fiscal stabilisation alone would lower mineral oil consumption, leading to reduced carbon consumption and nitrate proliferation. Based on this knowledge it is not defensible to allow wealth-induced $CO_2$ consumption to overburden and despoil our grandchildren's lives. Abetting such a prospect is witless; and, to the environmentally rational, obscene. Healthy human continuance depends on finding a way to bind us to adaptive reforms.**

---

Ernst Schumacher in his popular *Small is Beautiful* claimed, with 1970s optimism arising from the frothy sixties, that no viable business would ever squander its irreplaceable capital; yet that is precisely what businesses have done and continue to do; as with the fishing industry, the farming industry, the logging industry, the oil and mining industry, and the financial provisioning industry. Indeed my working time as a development planner has been largely directed at the enhancement of growth for poorer nations through an exploitation of their resource capital — resulting in degrees of failure I am now able to construe as a contra-success!

A personal turning-point came in the mid-1980s with a socio-economic-environmental project centred on the Dal Lakes in Kashmir;

a region-wide context which worked out win-win for community improvement and the environment. With this project, along with some other marginally successful work in Oceania, my challenge changed course to 'development with conservation'.[97]

Within democratic, pseudo-democratic and command administrations, the economy 'now' still trumps human low-entropy wellbeing for the 'future'. And so, far from inducing a withdrawal-from or a decrease-in the consumption of irreplaceable resource capital, there is a continued looting of the remainder pickings — fossil fuels, minerals, fish stocks, forests and aquifers.[98]

Now is the time to engage the acumen of inventive business logic — all those MBAs — to replace counterfeit economics with morally grounded and reality-led financial management which has an appreciation of the Second Law of Thermodynamics, the energy-entropy syndrome that transforms fossil fuels into stifling waste.

---

**Given the systemic 'financialisation' of national and international monetary arrangements, the challenge is to halt this mathematically contrived, mostly incomprehensible, and socially devastating manipulation.**

---

In the breech, for review, pull-down, and regulation are: Derivatives, Securitisation, Stock Option Dealing, Commodity Buyouts, Price Fixing Cartels, Franchise Scams, Futures Trading, Insider Trading, Tax Havening, Speculative Carry-Trading, Money Laundering, Buy-Out Mergers, Limited Liability Shells And Leveraged Dealings. There are serious failings and injustices about all the items in this listing, with 'tax havening' (at between $20 and $30 trillion) the most abominable.[99]

From an environmental perspective a continuing concern within nations of wealth is the fashion for the political 'left' and 'right' to both adhere to the growth-on-growth mantra. Indeed, when a growth indicator is down — this year not, say, a gain on last year — unemployment rises with the political opposition baying for a return to growth. And so, through the political medium (essentially our walleted vote) we still elect on the basis of desires rather than necessities. Societies have yet to cultivate a form of adaptation that alerts people to vote and sue in a supportive way for their grandchildren.

At a local *level* in most nations, legislation protects designated habitats and conserves some life support resources; prosecuting those who plunder or antisocially despoil the more precious patches of our environment. Regional *level* within-nation environmental criminality — toxic effusions, resource plunder — is equivocally prosecuted in only a few jurisdictions. There are, however, no effective legal mechanisms, or a World Court for Environmental Justice, available for the international *(macro-level)* prosecution of ecocide, the crime of wounding Mother Earth, the fifth humanitarian Crime Against Peace.[100]

Personal remembrance for most adults reaches as far back as our grandparent's adulthood; and includes, for some of us, our grandchildren's childhood; the normative continuum. Over this brief span of personal history our concentration on material growth has become dominant and all consuming.

Resource conservation and environmental health has taken a back seat relative to the expansion of capital, income increase, and an adherence to the 'wisdom of the markets'. This circumstance defines the Catch-21 bind of this century, between growth-for-consumption over the short term and conservation-for-survival over the longer term.

The economic-ecologic crisis of our time calls for a cultural innovation (bio-considerate, socially-driven community-based and largely prescriptive) with jurisdictions centering policies on economic fixedness, entropy slowdown and demographic stasis.

*Part B: Actions & Adaptations* (particularly Chapter 11 on 'Economics and Demographics') argues for an ethically and morally situated monetary balance.

## FURTHER READING

James, *Affluenza*.
Barber, *Consumed*.
Hart, *Capitalism at the Crossroads*.
Bower, *The Squeeze*.
Harvey, *The Enigma of Capital*.
Chang, *23 Things They Don't Tell You About Capitalism*.

The complexities of contemporary capitalist chicanery are examined, with wry humour and much indignation, in John Lanchester's *Whoops!*

# 8

# PART A: OVERVIEW
# LIMITS OF RESILIENCE

Humankind has dominated and altered the planetary landscape over a pinprick of time, only recently exceeding the biosphere's absorptive limits. And were this recent pattern of change to continue along the trajectory of the last fifty years — print money supply, carbon plunder, nitrate manufacture, and reproductive persistence — we knowingly pass-up the opportunity to gift a wholesome habitat to the next generation.

An expansion of dirty-carbon effusions after Peak Oil will exacerbate climate change; technically, when carbon gas concentration arising from the consumption of available and affordable coal and shale 'spooks' at around +4°C above year 2000 levels (tracking toward 2050!). Such an increase will trigger a significant tipping-point in human history; an event that will lead to an eventual meltdown of the Polar and Greenland ice fields and release vast stores of methane from the Arctic permafrost. The stakes are very high![101]

The wondrous biospheric system evolved by nature and the alluring economic system invented by humankind are colliding, a perfect storm. The main impediment to climate stability is the recent (1970s to 2008 and now re-evolving) fiscal transmutation that leverages consumer demand, increases mineral carbon uptake, and boosts population growth.

Perversely, much of modern living leads to an inversion of the very 'happiness' people seek from their surfeit consumption; inducing bouts of stress, anxiety, boredom, frustration, alienation, trauma, criminality, rage and gloom. Personal wealth and personal happiness are frequently misaligned, amounting to a phantasm.[102]

For the majority of us, family is the core to our being; especially the peripherals observed from the beginning and the end of middle-life, those lovingly tended grandparents and fondly enjoyed grandchildren. Virtuous existence is about conserving the wholesome habitat we inherited, and passing this on as a sustainable legacy. This is an assimilable notion. Virtue based on Spinozan logic 'Man Thinks' urges modern women and men to spurn seductive advertising, question the mantra of self-interest, and reject fallible edicts.[103]

Environmental virtue provides an ethical construct; engaging us to sue for a healthy habitat for our children. Such optimism needs to be tempered by the fact that consumer behaviours are entrenched in torpor and transitions take time.

**Three Civilisations: Two Failures**
**Humanity and the Biosphere**
**Consumer Culture: An Unsustainable Juggernaut**
**Environmental Awareness: Politics of Greenwash**
**People and the Environment**
**Technology and the Environment**
**Biospheric Economics**

The *Limits of Resilience* chapters, listed to the left, reviewed the influences on climate change through human agency.

Repeating that question first raised in the Introduction 'What do we now mean by progress?' The pragmatic response, in terms of human meaning, is to avoid extreme climate change and keep our habitat in good repair by holding to the sustainable ethic.

---

**HERE, THEN, IS THE DEAL:**

Keep a weather eye on the boundaries of biospheric resilience — the Part A evocations just concluded — and both prevent and adapt to change.

This involves advancing the sustainable ideal using every available policy provocation, prescriptive innovation and environmental defence — The Part B: 'Actions & Adaptations' which follow.

---

# PART B
## ACTIONS & ADAPTATIONS

"Life can only be understood backwards;
but it must be lived forwards."

— SOREN KIERKEGAARD

**Part B: 'Actions & Adaptations'** identifies the essential preventions, actions and adaptations for living sustainably within the limitations of planetary resilience.

Appendix C provides 'Exemplars from Oceania'.

# 9

# NEW PARADIGM:
## New Norms New Politics

---

**Preventions, Actions and Adaptations | Institutional Accord | Shifting the Political Paradigm | Community Engagement**

---

Those who take all they can from earth-bound resources on a greed-is-good basis damage the biosphere and downgrade our chance to secure a wholesome continuity. Other individuals, through necessity (their poverty), or consciously (their thoughtfulness), limit their uptake of earthbound and biospheric resources for the future common good.

What may be a surprise for some is that the overall ratio of 'plunderers' to 'conservationists' is a favourable one-to-four. We have got to the situation where the now wealthy one-billion-plus continue to despoil the future for themselves, dragging the five-billion of low to modest wealth, the Majority World, with them into an environmental quagmire. This highlights a lack of control over consumption and discard, and the delusion of expanding choice paraded as progress. The focus of attention in these 'Action and Adaptation' chapters is primarily on downsizing the resource consumption and waste discard activities of individuals and single-generation households in wealthier communities.[104]

Part A identified the Limits of Planetary Resilience, supplemented in Appendix B with parables, the Warnings From Oceania. Together with the Easter Island story (Chapter 1) these illustrate the disastrous consequence of resource depletion, overcrowding, external dependency, faulty technology, toxic pollution and flawed governance. None of these factors is of cosmological consequence; 'nature' simply does not 'care'. They are solely of concern for us and our co-inhabitants on planet Earth, the tiny blue orb in black space that both defines and incarcerates our frailty.

Sustaining global continuity — containing global temperature rise within +4°C of the Millennium Date baseline — is a complex matter of evolutionary foresight easily comprehended as a goal, difficult to attain as a target, and a serious challenge to hard-wire as political orthodoxy.

---

### Appendix C — Exemplars From Oceania I
### Niue: Benign Governance

---

Environmentally unstable behaviour, the pursuit of profit by captains of industry, and the profligacy of moneyed individuals and households (the 'weak democracy strong market' model), leaves us sleepwalking toward excessive and irreversible climate changes. Uncorrected, this institutional degeneration will go on to wreck personal lives and community wellbeing. In large measure it amounts to a *de facto* political failure.

## LOVE POWER AND DISTRACTION

From the mid-1990s on into the time of the monetary collapse of Indonesia, Thailand and Korea, financial conservatives failed to connect with a US President distracted from matters of state by efforts to cover-up his liaison with Monica Lewinski.

Joseph Stiglitz (2001 Nobelist) has admitted that he was not able to get to his otherwise socially democratic President. He was not able to urge action be taken against the protection rackets being run by self-serving bankers and hedge fund operators – to get the President to intervene politically.

Ten years previously, on a much smaller scale, David Lange, the Prime Minister of New Zealand, was embroiled in a liaison through which, and from which, he retained sufficient astuteness to terminate the fiscal waywardness of his Finance Minister.

The important point here is that even if political leaders in democracies have difficulty with understanding how money works, economic destiny must be managed and patrolled through due political process in the interest of society as a whole.

A useful reference is Christensen's *The Innovator's Dilemma* (1997).

The impediments to progressive change include blinkered vision, antediluvian protocols, skewed markets, skimpy punishment systems, turgid bureaucracies, and a widespread political obsession with growth.

Other obstacles are the components of social organisation: hierarchy, ideological beliefs, race conflict, gender inequality, envy, hubris, and vanity. All these constrain peripheral and longer-term vision, and generate a fear of backlash; driving politicians to give us what we want now, continued consumption, rather than inducing us to strive for wholesome survival in the future.

To this confusion add venality; the high proportion of politicians capable of wrongdoing and wrongheadedness beyond mere procrastination; embezzling occasionally, acting meanly when there is an opportunity for petty gain without getting nabbed, being foolhardy frequently, and seeking to maintain political capital by suckering their electorate into keeping them in office. The socio-environmental culture argued for in these pages involves the alignment of corporate management and political interests, 'left' and 'right', with the longer-term health of our habitat. It is a 'public interest first' credo bundled-up with economic stability, resource use constraint, sensitively managed fertility decline, and environmental justice.[105]

Hyper growth and consumption lowers our gaze from distant horizons to the immediate foreground, avoiding an ominous future in preference to a hedonistic present. This leaves little room for us or our politicians to take stock throughout days chock-full of clarion calls, diversionary tasks, sound-bite grabs and exhortations to earn-spend-consume-discard at an ever-accelerating velocity.[106]

Excess consumption, the fault, has junk advertising as its muse; inducing multiple lifestyle choices, gadget fatigue and overload dumping. Denying reflective space to reason ways to live lives of environmental virtue has infantilised politics, inducing voters to support those leaders who vow to keep growth on a fossil-fuelled path.

---

## CONSIDER THESE APHORISMS:

"We are so much absorbed in the toil of the day that we leave the morrow to take care of itself."  LORD CURZON

"Human beings only pose problems they can solve."  KARL MARX

"People cannot stand too much reality."  CARL JUNG

---

Reflecting these thoughts against the pressures politicians are under to get re-elected enables us partly to comprehend why it is that they tend to opt for the feel-good-now option rather than the maintain-good-later alternative. The situation requires overall partisan agreement on environmental policy — a new social democracy — in order to ensure that climate stabilising objectives are not compromised by economic folly.

Some Creationists claim that humankind emerged in ready-to-roll form, 6022 years ago. Cranked, as this may seem, wry amusement can be drawn from the fact that human groups did indeed cohere and form settlements about that time. From around that supposed Date of Creation larger city-state settlements, beginning with Sumer, were forming. Likewise in ancient China and Central-Southern America stable and sizeable nation-states emerged; and from Ancient Greece a collection of democratically organised entities flourished and consolidated.[107]

Now about half the world's people are governed by democratically elected growth-driven urban-dominated administrations. Although flawed, democratic governance along with expansive corporatism and business growth, remains in place as the favoured (least-bad!) arrangement. Fortuitously, a new dimension arises in that electronic ubiquity now facilitates a dialogue, of sorts, between people who may never meet each other in the flesh. This should make it easier for governments and interest groups to test and receive opinion and debate principles; and air differences and come to agreement about demographics, economics, carbon use, nitrogen fixing and abrupt climate change.

Fundamentally, securing the future for our children's children is related to finding ways of reducing mineral carbon mining and nitrate manufacture. Much impressive knowledge, substitute technology and operating skill is available. The sticking point is our inability thus far to fire-up political respect for environmental limits, and commit to resource-use constraint. We are in need of a sustainment paradigm that repudiates econometric growth in favour of fiscal stability.

The survival strictures that emerge from understanding the Limits to Planetary Resilience are 'think sharp, get wise, slow down', the grandparent-to-grandchild challenge. This is the 'smart' (clever, wise, sagacious) stability position to the current 'slick' (tricky, devious, shifty) growth situation. It constitutes a challenge for jurisdictions seeking to find a way back, via democratic governance and the rule of law, to something like the levels of resource uptake, which produced the moderately degraded landscapes of the 19th and early 20th century. It is not an advocacy for killjoy pessimism; but at its core it does involve a massive reduction of carbon uptake and nitrate manufacture. There is another benefit; the satisfaction of maintaining a personal balance with nature; sustaining your home-patch harmony.

---

'Continuity' and the attainment of a 'sustainable stasis' is our social responsibility: it also identifies an institutional, community, corporate, and political obligation: a salient position which recognises that our earthly orb is the sole and only 'bell jar' available for future human occupancy.

---

The political challenge relating to climate stabilisation also requires a style of social democracy that identifies and respects the environmental limits of resilience. The emphasis is on the capacity to reason and review options for prevention, action and adaptation. It is grounded in an appreciation of kinship, seasonal rhythms, and the bounty of nature. To this end a store of managerial experience and knowledge has been accumulated, out of which democratically elected government is an impressive achievement.

Beyond the political challenge in jurisdictions of wealth, is the challenge to spread the fossil fuel and agricultural-nitrate cutback message to less wealthy jurisdictions. It is essential to have all the major players within the democratic tent, sharing findings, the free press, and the rule of law. Only in a globally concerted way can regulated money supply, coupled to sensitively coerced population stabilisation, be effective in the pursuit and attainment of reduced carbon and nitrogen uptake.[108]

Pan-political agreement and concerted adaptation is the key to halting humanly induced climate change. And the nations best placed to initiate stabilisation and instil action are the democratic enclaves of established wealth, which happen to also be the major part of the problem. If the nations of wealth do not make progress with a reduction of carbon fuel emissions, what can be expected from the democracies of the subcontinent, the remaining Central Command economies, and members of the OPEC cartel? Climate control has to be fashioned on a pan-political, socially democratic, platform. This requires that the new politics defer to adaptive realism; fashioning, endorsing, and implanting cuts in carbon use and nitrate fixing for the longer term interest of everybody everywhere. A useful contribution to the 'new' political rubric has been provided by Anthony Giddens (2009) and a selection from his findings has been built into the next diagram below.[109]

## SHIFTING THE POLITICAL PARADIGM

- Promote sustainability as a creative human right
- Fashion a left-with-right sustainable radicalism
- Strengthen democratic institutions within societies
- Motivate for an ensuring, enabler, facilitating state
- Pursue environmental justice with a hard edge
- Converge: Politics, Science, Economics
- Think ahead: Foretell risks and uncertainties
- Fashion 'assessment' and 'forward planning' protocols
- Maintain rustic resilience and hi-tech gains
- Include non-human organisms as moral referents
- Emphasise incentives as well as restrictions
- Legislation: Mostly enabling, but also restrictive
- Enforce 'boundaries' and the 'polluter pays principle'
- Regulate counter-environment businesses and markets
- Support low-carbon uptake and use practices

There is also the partly political matter of establishing how, in practice, we can go about adapting to climate change. The response — myriad suggestions, ranging from the micro (shopping with a string bag?) through to the macro (carbon constrained manufacturing and farming?) — is well served by the shelves of literature listed in the References. The problem is that within this list of writings insufficient progress is registered on the vital matter of identifying and acknowledging the limits of resilience and the specifics of action and adaptation. The recurring climate change conferencing of the past has not been able to overcome this obstacle. It is a theme the later chapters on 'Conservancy' 'Goals' and 'Targets' confront and address.

The 'new' new political objective seeks also to enshrine in policy and law a non-market edict for halving mineral-fuel and manufactured-nitrate supply. This, coupled to a doubled-up efficiency of energy utilisation, steady state economics, fertility decline, and reforestation would induce a reversion to climate stability.

Within this mix it is the current lesser-proportion of wealthier people — the global fifth — who must adopt the sustainability ethos person-by-person, community-by-community, institution-by- institution, business-by-business, and nation-by-nation.

Politicians sabotage the future of humankind when they bind themselves to advisors for growth and consumption. Recall, as an example, the political astuteness of JFK, besieged by belligerent 'attack advice' during the Cuban Crisis, taking the 'right brain' withholding decision that secured a future for American Society.

## COMMUNITY BY COMMUNITY

Elinor Ostrom shared the 2009 Nobel Economics Prize for work which established that community co-operation for the common good can overcome individualistic and corporate driven despoliation of natural resources; notably so within economically challenged societies.

Her work also aligns with the principle of subsidiarity which seeks to engage local communities to manage and conserve local resources through local creativity and understanding, and the use of traditional (local) penalties.

What is required of the collective political conscience is entry into a moral realm that encourages them, representing us, to secure an environmentally healthy future for all those who inherit. To this end there is a need to expose the falsity and establish the hopelessness of the existing growth-market nirvana; for nowhere in that maze is there enduring substance. The faith required to conserve the future habitat for our grandchildren is not an either-or alternative; in terms of Thomas Kuhn's *Structure of Scientific Revolution*, it amounts to a paradigm shift.

In that curious role reversal previously mentioned, only the poorest nations start from pole position with regard to coping with climate change. In contrast, nations within the OECD-OPEC cluster, along with the rapidly expanding economies of China, India, Vietnam and Brazil, remain transfixed by the Messianic attraction of money; and have little understanding of how to get from an environmentally endangered, econometrically destructive and socially imperilled 'here', to a future-proofed 'there'. Thus far the nations of wealth, and people of wealth, have only a superficial and oftentimes whacky green credo to lead them out of harm's way.[110]

---

## ADAPTATION IS THE TASK BEFORE GOVERNMENTS

- To act morally, decently and assertively as the promoter and arbiter of sustainable necessity
- To put environmental protocols and preventions onto institutions, communities and corporations
- To shape mindsets beyond self-serving individualism toward the common good
- To move swiftly for fiscal stability, and stealthily toward demographic stability
- To fashion communitarian, pro-sustainable, socially democratic institutions
- To pledge environmental justice for our offspring and ourselves.[111]

---

Unfortunately *both* centre-left and centre-right representatives, *both* in attendance at climate-adjusting events, have *both* got trapped into the output of anodyne, unwieldy and empty proclamations, which is tantamount to a human travesty.

Voters elect those they believe in. The call now is for the electorate to embed in politicians a drive to heed and provide for the longer-term needs of their electors, and to overcome previous short-term growth-policy failings. The challenge for politicians is to grow institutions for both the conservancy of climate and constituency.

Hovering around the GlobalGood proposition, advocated in the next chapter, is the matter of what is behaviourally appropriate. In the past Homo sapiens, the only specie with a known cognitive purpose to life, set a major part of non-domestic intent at 'enslave expand exploit exhaust empower'. Despite large swathes of irreversible environmental harm and resource depletion, that pattern remains imprinted, installed and perpetuated. There is a need for a global seismic jolt of awareness about climate change on the part of individuals, households, businesses, communities, institutions, corporations and governments.[112]

Philosophical pundits urge us to 'live well and be good' as though putting such empathy out there amounts to an operational principle with potential. For climate change deniers specifically, and most of us generally, the 'be good' exhortation falls foul of the magnetism of 'self interest first'.

Looking-in as it were from off-planet gives rise to an objective question: 'does extended human wellbeing and planetary degradation really matter?' One position is that a clear majority of the now mainly agnostic better-off do not believe that resource plunder, toxic waste discard and environmental degradation (let alone poor people's famine and pestilence) constitutes wrongness, or need be of concern. To many, what is postulated about living sustainably suffers from individual indifference, remains philosophically fallible, and is of marginal interest politically.[113]

My position, in line with Tony Judt's *Ill Fares The Land*, is that "We need to rediscover how to talk about change: and how to imagine very different arrangements".

In this we had better be pretty quick, for the probability is that by the middle of this 21st century 'artificial intelligence' will eclipse and in many contexts supplant the efficiency of human decision-making. If artificial intelligence is programmed, or instructs itself to operate for our benevolence and wholesome continuity, well and good; but what if it is programmed to pursue mayhem by initiating the likes of cyber sabotage invoking widespread utilities failure?

## FURTHER READING

United Nations, *Resilient People: Resilient Planet.*
Amartya Sen, *The Idea of Justice.*
Karl Popper, *The Open Society.*
Havel, *Art of the Impossible.*
Bok, *Politics of Happiness.*
McDonough & Braungart, *Cradle to Cradle.*

# 10

# INFORMATION & EDUCATION: The GlobalGood Curriculum

---

Environmental Rights | Environmental Justice | The GlobalGood Protocol | Outer-Space and Cyberspace Surveillance | Transparency and Disclosure | Adapting to Change | Curricula and Pedagogy

---

An increase in temperature beyond two more degrees (four degrees above the Millennium Date level) will trip an irreversible switch. Ice fields will melt, low lying coastal lands will be inundated, savannahs will crisp to toast, tropical rainforests will turn to savannah, temperate zone forest fires will rage, species will be wiped-out. These extreme events now have a name 'climatic weirding'; thematically evident as fictionalised disaster-reality in Margaret Atwood's *Oryx and Crake* (2003); and Cormac McCarthy's *The Road* (2006).

Concern for the resilience of the biosphere, the parameters of a sustainable culture, and the future for humankind crowd in.

- What will happen to pump prices beyond Peak Oil?
- Will carbon consumption increase using dirtier sources?
- Will the smaller nation states do better than the large?
- How difficult will it be for large conglomerations to cohere over policies?
- Will some larger states disintegrate?
- Will some cities and densely populated regions descend into anarchy?

- Will urban populations turn back to the land?
- Can a decline in human population and carbon-usage be managed in humanistic ways?
- Might the poor become the most capable survivors?
- Is bottom-up more effective than top-down?
- Will 'spoiled child' forms of democracy prevail?

Understanding climate change, contributory causes to this change, trends in the pattern of change, and the planetary boundaries of resilience, are essential fact-based data. In this context fact-based is unbiased and tamper-free explanations of *'What is happening?'* along with margin-of-error extrapolations of *'What could happen'*? It is something more proactive and driven than the IPCC process, useful though that has been.

In this age it is possible to collate vast amounts of data, do massive calculations, predict multiple scenarios; and render information and findings accessible in every language world-wide via wireless, print, television and otherwise electronically.

---

### Appendix C — Exemplars From Oceania II
### Natongandravu: Lessons From Experience

---

The flow of information about weather spike phenomena is a daily constant. News of unprecedented floods, droughts, and tornadoes is in the print media (even the gossip magazines), in television bites, in learned articles, and most immediately on radio, the web and by cellphone; all indicative of climate change awareness. But when it comes to the matter of boundaries (limits) the message is not getting through. Obfuscation of and a fatalism toward risk is evident; for it is still not widely perceived to be in our collective interest to introduce Plan B, when it is not yet acknowledged that Plan A isn't working.[114]

A chilling angle to the matter of obfuscation is that policy makers in big business and government largely ignore unwelcome statistical evidence. This includes the fact that further deep-sea oil 'harvesting' along with the mining of dirty coal and tar sands will release and produce unassimilable amounts of $CO_2$, nitrous oxide and toxic ozone.

Strategists tripped-up by the complexities and frustrations of reality often say that they understand the facts and agree the conclusions, yet still they mostly fail to pursue adaptive reforms. Compounding this problem: we settle for short-term acceptance based on personal hunch and folkloric conviction that 'the future will take care of itself' mainly because, right now, fossil carbon energy is compact, cheap, portable and available.[115]

Negatively geared plans are seen as a loser's credo from the perspective of the wealthy; those previously able to create their own reality, summer goods in the supermarket all year round! They have been led on by low-cost carbon supply, coupled to no-cost carbon gas discard. Up until now this opportunistic 'success', when locked into short-term memory, has cast government and corporation leaders as the progressive good guys, and the environmental realists as weak and wishy-washy.

Hard facts about biospheric composition ($CO_2$ in the atmosphere, soluble nitrates in the hydrosphere) needs to be better ordered packaged and disseminated, with open access to analysis prognosis and peer reviews. In truth the mental make-up for atomistic people of wealth is dominated more by a hole burnt in their pocketbook than a million hectare hole burnt in the Amazon Rainforest. It is not yet in the interest of corporations doing what they are inherently ordained to do, or most government's inherent drive for electoral survival, to take effective action against climate change, much less formulate Plan B.[116]

As noted earlier, there is much to admire in the work of the IPCC, particularly the stand taken in relation to influential contrarians and moneyed lobbyists. Yet although the climate change challenge translates into a social challenge, it has never been the Panel's job to engage with the political process. They do not seek to engineer adjustment because this is not part of their 'stolid' information-based brief; earlier reports coming through as 'all caboose and no engine'. They were and are focused on discovery; they report climate change findings of relevance to the future.[117]

In contrast to the IPCC endeavours, the growth proponents are 'slick'. They are proactive about lucrative deals and investments now, seeking to leave the climate damaging issue where it has always languished, in the 'too hard pending tray'. They guzzle imported Perrier today, and are indisposed to discuss the absence of potable tap water tomorrow. They access all available hedonism to hand, and overlook the case for enduring environmental health. They enjoy petrol-powered mobility and loath naturally intended perambulation. They register an electronic clip, but spurn polemic discourse.

Environmental good health is a social need which involves an understanding of the adverse indicators — the 'how come?' problematics, and the 'and next?' socio-political options — and to come up with behavioural guidelines that inform and educate to this end. Holding out for a normative past-present-future continuum amounts to reality denial.

Secret knowledge and secret decision making has always been a feature of transnational corporate management. To a policy director knowing 'what's-up?' exclusively, represents power, power to profit. To the rest of us, informing and educating 'what's up?' is inclusively progressive. Honestly brokered knowledge openly informs and facilitates the attainment of social goals. The catchwords are: **inform, learn, reason, prevent, adapt, orchestrate,** and **deliver.**[118]

---

A GlobalGood data assembly, analysis and prognosis programme has a pivotal role; matching opinion against data, inference against scientific fact, and allowing individuals and communities to adjust their ethical posture in response to real-time findings and future limits.[119]

---

Two areas of concern, profiled previously as largely unchallenged, are future population expansion and hyper money supply. Unchanged they work together, whatever the numbers, to accelerate climate change. Neither corporations nor governments seek an open dialogue on constrained growth, or want to declare positions relative to reductionist demographic and fiscal goals.

There is a reluctance to support research addressing the impact of mass population (fearing the eugenic backlash), or the impact of mass monetary blowback (fearing economic panic). Within the mix it would be unbalanced to single out population-lowering as the principle agent for arresting climate change because the core to reducing consumer excess arises mainly from that lesser part of the population which holds sheaves of wealth; being also the cohort with longer survival rates.[120]

Policies to reduce global population, working up from the very poor, would have a slow beneficial effect on climate change. On the other hand a stabilisation of hyper-inflated wealth, and carbon-and-nitrate consumption, would have an immediate beneficial impact. The position arrived at here (elaborated in the next chapter) is that in terms of biospheric resilience the admittedly improbable but hopeful priority is monetary stabilisation leading, as a matter of course, to a localisation of lifestyles and a lessening of carbon consumption and nitrogen fixing.

Fiscal meltdown was inevitable both at the time of the 1929 crash and the comparable mid-2008 blowout. Now, through information and learning, the miasma of previously blanked-out information about collapsing investment banks and reinsurance scams comes to light.

> **The point being established is that climate stabilisation can only be congruent with an axing of money supply expansion; calling for regulation, transparency, honest dealing, disclosure, and all transactions put through the books. Spotlighting the stabilisation of economic growth, involving a poleaxing of all forms of faux wealth generation, is the most necessary and least considered contributor to biospheric stability.[121]**

Fired by turbo-wealth, climate change accelerates. Diversionary information and undisclosed data obscures the necessary avoidance of further global temperature rises by two more degrees (+4° above the Millennium Date level). The challenge is to find a way to return to a healthy eco-stasis, living off the global living resource 'interest' within our solar energised biosphere; giving up on the drive to plunder the once-in-history carbon and mineral 'capital' at a health-damaging rate.[122]

An issue to sheet home about eco-environmental education (this chapter's theme) is that poignant historical lessons have been mimicked and illustrated in the near past through the likes of the *Appendix B: 'Warnings from Oceania'*: the disastrous massing of population on Tarawa; the calamitous fiscal flow and ebb on Nauru; the institutional weakening induced through outsider dependency in American Samoa; cargo-cult delusion in the Solomon Islands; resource loss on Easter Island; and the disappearance of usable land in Tuvalu. These are micro-examples drawn to the attention of us all within our already destabilised global bell jar.

Our failing? In the previously quoted words of Curzon, overwhelmed in 1909 by the immensity and complexity of the Indian Subcontinent, we now, over a hundred years later, tend still to "…leave the morrow to take care of itself" haunted by the stalking horse of disinformation, indecision, self-interest, apathy and immediacy.

The people I connect with during my daily round divide between wealthy multiple-property millionaires, green leaning semi-rural preppers, and busy-busy townsfolk; all with a mixed bag attitude as to how they would cope in a climate constrained future. One rich guy has loads of food and fuel stored for his family 'against the day'. My greenie friend wears homespun, eats homegrown, has heaps of solar powered top-shelf electronics and helps to homeschool her grandchildren. And my typical workfolker pal and his blended family go for broke with a double mortgage, a pair of utility vehicles, and a culinary preference for meat, beer and ice cream.

We cannot, *not* be supplied information, *not* be educated about the facts and process of climate change, and *not* have access to techno-fix opportunities. In this the electronic transmission of information is a blessing; but although the digital 'library' is cheap it includes a ragbag of irrelevancies way off target. Electronic communication bridges distance but mostly lacks prescience.

For all of us the intrinsic meaning and purpose to life is its perpetuation in good health. We 'control' so in effect 'own' the planetary wonderment. We consume biospheric and earth crust resources, and in this way fashion what is widely perceived to be human advance. But now we fast approach that reality fork where a future gets fashioned which can be either a morbidly degraded inheritance, or a habitable continuum.

Talk about the excitement of learning? The whole of humankind, but particularly the well-heeled one-fifth, are in a 'school of environmental studies' where the only way to 'pass' the upcoming 'exam' is to make the correct responses. This is a programme invoking adaptation to climate change on several fronts, seeking movement on matters as diverse as ethical sanctity and pragmatic cleverness. The exam-style rubric runs: get the vitally important answers on climatic stabilisation wrong, and you are on your way. Sadly, there is no retake.

Install and update through international provisioning (UNEP, WTO, IPCC, UNFCCC) an open-input, free-access, unfettered GlobalGood facility to research receive collate interpret trend forecast update and disseminate — openly in the public domain — all manner of demographic and economic shifts and resource information influencing or foretelling future biosystemic change.

At a recent academic *viva* a worthy presentation harnessed considerable computer resources to proscribe the diffusion of emergent vegetation in the windfall spaces within a hurricane-ravaged rainforest. The variables were considerable, the permutations massive, the complex determination convincing. I asked the supplicant about the feasibility of designing a programme into which all measurements of biospheric-change data from all parts of a region could be fed into a supercomputer to plot all manner of trends. The response was 'If statistical trends were sought a supercomputer would not be required, however one would be needed to extrapolate continuously updated forecasts imputing multiple variables from myriad sources.' In short: a GlobalGood information service is operationally feasible.

## SURVEILLANCE
## OUTER-SPACE AND CYBERSPACE

Between them, Nasa Earth Sciences and Google Geo have the potential to pin point the likes of unanticipated $CO_2$ effusions and rainforest clearances by capturing the raw data and making the findings available.

The satellite technology is proven but not funded; and without raw data from Nasa as a provider the Google Earth engine is impotent as an informer.

Supercomputers exist, mostly for military and strategic purposes, at the rate of about one to every five millions of OECD population, and now also in India and China. Aside from the fact that UNEP, Worldwatch, some Universities and a few State Departments are running in-house climate change programmes, there is a need for access to GlobalGood research and information which is independently funded and operated, conspicuously profiled, openly transparent, readily accessible and reader friendly.[123]

---

Key phrasings around the furthering of environmental justice through a GlobalGood facility include: **Operational Independence, Multilateral Funding, Research Enablement, Open Management**, and **Continuous Overview**.

---

A GlobalGood facility, operationally independent and constitutionally protected, is a response to the fact that the monitoring systems already in place are inadequately supported, under-profiled, fault-driven and procedurally hedged. Access to information on environmental, social and economic change, continuously upgraded and reforecast, lies at the heart of the GlobalGood recommendation.

GlobalGood represents a 'big idea' for facilitating the research and interpretation of the copious amounts of data required to sheet home understanding about planetary limits to change. In this the power of open-source software, possibly federating widely dispersed under-used computer capacity, can be pledged for GlobalGood use — a neat example of excess digital capacity being harvested to do socially good work.[124]

Providing factually unbiased information and analyses for decision makers at all levels of management, in effect those who need to know about the real facts and the real effects of climate change, is the driving purpose behind the GlobalGood enterprise. Such a service would also confront deniers, doubters and deceivers; those who need to come to terms with information not previously made available to them; serving-up environmental rights, discard limits, sustainable ethics and modes of prevention and adaptation to climate change.

---

Vital are pre-school, primary school, high school, craft school, college and university curricula and community forums infused with information about climatic change and the Big History limits of biospheric tolerance to resource exploitation, waste discard, and environmental degradation.

---

## FURTHER READING

CONTEMPORARY:

Thyer, *Grey World Green Heart.*

Suzuki & McConnell, *The Sacred Balance.*

Saul, *Equilibrium.*

Womac & Jones, *Lean Thinking.*

Benkler, *Wealth of Networks.*

Wilkinson & Prickett, *The Spirit Level.*


FROM THE PAST:

Rousseau, *The Social Contract.*

Perry, *Realms of Value.*

Illich, *Tools for Conviviality.*

Rexroth, *Communalism.*

# 11

# ECONOMICS & DEMOGRAPHICS: The GlobalGood Drivers

---

CORE IMPLEMENTATION OF: Fiscal Policy | Disengaging From Exponential Growth | Fiscal Stability | Greener Economics | Digital 'Gold' Standard | Revisiting the 1936 Chicago Plan

CORE ACTION ON: Population Policy | Fertility Limitation 'Two', 'One', 'None' | Inducing Socially Acceptable Fertility Decline

---

The three chapters following later— Resource Conservancy with Waste Conservancy, Goals, and Targets — are the policy and practice guidelines strapping down the core actions for achieving climate control. As nations go forward it has to be taken on board that the paradigm of growth and discard in the recent past will not extrapolate glibly into the future, for there are absolute waste absorption limits. This constitutes a certainty. The pivotal deal for a human future of worth is an adaptation where finance is stabilised to serve a resilient sized population within entropy-constrained parameters (Chapter 7).

Exhortations to turn off appliances on standby and to recycle packaging are important yet insubstantial, fig-leaf coverage from those of relative wealth with pinprick impact on curtailing the overall consumption of mineral carbons and industrial nitrates. In this chapter attention is focused on the bottom-line essentials and the must-do actions for reducing fiscal leverage and lessening population impact.[125]

---

Appendix C — Exemplars From Oceania III
Rewausau: Civilised & Sustainable

---

The mathematically contrived and inflationary racketeering, which led up to the 2008 financial meltdown induced four extrapolations over the previous twenty years; a doubling-plus of made-up money, a doubling of world population, a doubling of carbon dioxide emissions, and a doubling of the midpoint global temperature rise. Net result: the first decade of the New Millennium ended in quadruple disarray — fiscal chaos, waste excess, over population and irreversible climate change!

Wealth accumulation and wealth deployment is the key stoker for all four of the doublings and undoings. Here arises a question: is there in those individuals and jurisdictions of wealth a willingness to contribute to climate stabilisation from the use of their imputed worth and buying power? What, it can be posited, ought rich nations and rich individuals do with their container-sheds of money but invest in future climatic good health? The answer is, despite that necessity, the accumulators of wealth (excepting a few entropy conscious individuals) do not embrace biospheric rescue; their priority is to keep themselves financially secure.[126]

Turning to the matter of population. Were the seven billion human mass halved through constraint and attrition to say three billion, would that lead to improved climate stability? Possibly: provided the top billion or so of wealth were diverted from increasing their carbon and nitrate consumption. Knock-on problems arise. Strategies for reducing population raise awkward doubts. Reduction by birth limitation can only be attained slowly through the diffusion of education, enlightenment, advice and sensitive practical help.

---

**Population Number Reduction (Gradually), Wealth Impact Stability (Vitally), And Fossil Carbon Emissions Control (Urgently), are the core climate stabilising factors identified in Part A which defines the Limits of Resilience.[127]**

---

For most of us it would seem preposterous to prescribe a coupling of population-reduction and monetary stability; and to link this to a quartering of carbon consumption. Yet it is meek, defeatist and illogical to continue with a cash led carbon uptake that delivers a dead-hand stasis.[128]

Nations of wealth have fertility limitation and growth of population pretty much under voluntary control. In poorer nations the reduction of overall numbers is a more complex objective, giving rise to an argument for proscribing differentially (on the matter of fertility limitation) for rich and poor populations. This comes about because the bulk of the increased $CO_2$ over the last quarter of a century has occurred in, and comes from, the nations of wealth where population increases have been slight. The point is that an even lower rate of wealthy nation population growth on its own would not produce a lower rate of $CO_2$ emission; indeed the reverse, were unregulated money supply to increase. A slowly lowered rate of population growth in the World's poorer nations, if all the while they considerably increased their wealth and mobility, would also fail to reduce overall levels of greenhouse gas emission.[129]

Given widespread belief in an expansive global economic system, that system became the overall control mechanism. At first economic tinkering was introduced as an environmental adjuster; coupled to exhortations for individuals of wealth to behave better, make appropriate purchases, live more lightly, waste not and so on.

The softly-softly approach has not made a beneficial impact. And the extensions to tinkering, official Emissions Capping and Emissions Trading (Cap-and-Trade) along with other buy-out and swap variants, are not proven. These are the kinds of instrument which innovative entrepreneurs construe as a way for them to benefit, financially. Cap-and-Trade also fits pork barrel politics perfectly as a tool for favouring (in effect subsidising) selected industries, and for targeting marginal constituencies with benefit-bribes in order to stay in power.

Mildly pursued intervention — the thin-end of the economic wedge — manipulates economic behaviour with minor inducements and paltry penalties; care being taken to not dent private wealth or upset the political party in office. Over time, a tax-and-penalty economic wedge can thicken, although it will always be an 'adjuster' activity rather than bringing about an 'achiever' result.

The outcome sought, environmental equilibrium, eludes for two fiscally relevant reasons:

1. Price and tax tweaking makes little change to the damaging lifestyles of the one billion-plus people of greatest wealth who, as the main climate altering culprits, are able always to pay more to consume-and-forget the downside externalities they generate.

2. The poorer four billion-plus, driven to improve their lot by earning and consuming more, increase their rate of environmental degradation. The world at large requires a stable, reliable and constant trading and clearing house system; something like a return to a Gold Standard, without the gold, say as 'digital gold'.[130]

After liquidity decline and recession (early 2008) there was a brief slowdown in oil uptake, manufacturing output, and consumer splurging. Following this unintended environmental benefit (a consequence of wealth erosion) administrations should be drawn to see how they can support the biosphere and their offspring by giving up on the inflationary protection (socialising) of banking, corporate and high-end 'mistakes'.

Here arises a perverse twist and an irony: the evaporation of up to 40% of personal wealth for many persons of mega-wealth as a consequence of the 2007–08 economic slump, actually supported their case for a reliable and stable fiscal system.[131]

Retaining global temperature increase within the Year 2000 +4°C outer limit is an easily stated intention, up there with the deceptive 350–400ppm limit. But keeping within these easily trilled numbers happens to be crucial to a dignified human future. In this it has to be accepted that economic stabilisation and population reduction will be a variegated and tardy transition. Protocols have to be calibrated differentially between rich and poor nations; involving a mix of emissions-reducing, growth-quelling and population-limiting incentives and penalties.

## ECONOMIC STABILITY

Speculative monetary gain which societies fail to repress, or control, hurt us all. Bio-sensitive governance in combination with a stable non-speculative monetary service and no-growth demographics is the enabler of dignified human continuation.

---

Fiscal stability is imperative. Policy and practice is required which is secure and stable, and both real-time accountable as well as providing sanctions against a repeat of previous boom-and-bust mistakes. Government's must accept the responsibility to get in there and do the job that has to be done.

---

## ECONOMIC STABILITY & RESILIENCE

**MACRO ACTIONS:** [Refer also to Appendix D]
Initiate fiscal stability reform
Establish risk oversight agencies
Install 'transaction' taxes (Tobin Tax)
Impose exemplary eco-harm penalties
Dealings: transparency, honesty, openness
Separate 'utility' and 'investment' banking
Substantially upgrade 2010 Basle III rules
Apply supercharged penalties for cleanup
Critically review carbon futures trading
Allow 'green' exchange systems
Surveillance: checks balances and feedback

**CORRECTING ACTIONS:**
Blitz shells, scams, flash trading
Regulate and discipline political lobbying
Encourage gender harmonised governance
Prohibit off-balance-sheet activities
Hobble options and futures trading
Discipline securitisation
Outlaw credit default swapping
Restrain commodity buy-outs
Suppress 'secret' and 'shadow' banking
Deny harmful-to-climate subsidies
Surcharge high-end consumers

**CONTROL ACTIONS OVER:**
Hedge Fund trades and franchise scams
Transaction-fee bonus schemes
Insider trading and political bribery
Institutional insider trading
Low cover (sub-prime) lending
Tax havening and money laundering
Profit-absorbing structural transactions
Speculative carry trading
Buy-out mergers
Golden parachuting
Dividend protection scams
Risk incentive contracting
Indefensible, unmerited, pay and bonuses

The Macro Actions listing, about which I claim only derivative understanding, have been sourced largely from articles by Stephanie Blankenburg, Gabriel Palmer and Robert Wade, in the *Cambridge Journal of Economics*, July 2009. Reference also to Joseph Stiglitz *'Moving Beyond Market Fundamentalism to a More Balanced Economy'* 2009.

Antisocial and quasi-illegal as well as illegal enrichment scams must be poleaxed. Speculation 'gone wrong' must never again have the correction of such mistakes 'socialised' from the public purse. A principal objective is adherence to a negative. This involves avoiding a return to slight-of-hand liquidity, debt burdening and hyperinflation as the prelude to blowout.

Monetary services were never intended to become government sanctioned 'casinos' allowing insider game-of-chance rewards and absolved losses to a few well-placed 'players'. Now, more than at any other time in monetary history, it has become crucial to establish a stable facilitation of trade and exchange under a security-certain arrangement. The style of fiscal guardianship required has to be stolidly formulaic, much more so than 'green economics' which doesn't seek authoritatively to amend the money-multiplier system.[132]

Sanctioned by the IMF, Jaromir Benes and Michael Kumhof have 'revisited' Irving Fisher's 1936 'Chicago Plan' for establishing a solely state-backed 'sovereign' money supply service using government issued credit, somewhat in the way most of us believe, naively, contemporary money supply works.

---

**The 'Chicago Plan Revisited' and the Sovereign Money Movement would put a stop to bank-led inflation fuelled with bank-created credit by requiring banks to act as intermediaries, passing-on government issued money to borrowers. By legislative command, public debt would be eliminated, bankers dethroned, the credit-cycle bumps eliminated, and money supply stabilised.**

---

The lesson learnt from the near past is that open access to easy money cranks-up to consumer excess, overhanging debt, then debt default. This works through as both environmentally reckless and socially damaging. No danger arises from regulation. Much better to have prescriptive controls than successive blasts of monetary expansion-implosion, with calamitous consequences. Keep in mind that post-Enron only a few Madoff-style really nasty crooks, and equally few scammers from previous positions in investment banking or insurance, are doing serious jail time for theft, fraud or malfeasance![133]

The *laissez faire* expansion of economic imperialism and money supply over the past forty years has damaged the environment, ruptured economic order and wrought social havoc; leaving our offspring hostages to enormous debt. A new style of monetary governance is called for which provides a transparent non-speculative utility for society, somewhat in line with the participatory economics being advocated as Parecon and the already mentioned 1936 Chicago Plan Revisited and Sovereign Money formats.[134]

The Macro, Correcting and Control lists given in the derivative but credible Economic Stability and Resilience aside, are linked into the practical National-Local-Personal 'markers' given later (Chapter 13).[135]

There is something else. Tertiary education mostly instructs and then supports graduates in the arts humanities and sciences for lifetimes of broadly social service. Other academic programmes offer training for the practice of medicine, planning, architecture, law and engineering; after which the meritorious are fee-rewarded and honoured by their peers for their specialised expertise. Ethical canons and practice codes, both mandatory and elective, inform and guide the conduct of these practitioners in the delivery of their skills.

In contrast tertiary courses in financial management (now morphed into 'business schools') instruct tacitly (therefore outrageously) for individual enrichment, shameless risk taking, resource plundering and revolving-door opportunism. The installation of this rottenness can be laid at the portals of academe. Failure to provide and apply an 'ethical compass' for those in the financial services sector is ethically, politically, and therefore socially, irresponsible.[136]

---

**In terms of culturally responsible policy it is essential to ensure that people in finance are engaged on a society-serving basis; putting an end to hot money profits and hidden debts. The common identifier in professional employment of all kinds has to be hallmarked 'socially useful'; with citizens in the private as well as the public workforce accorded community recognition for their public service.**

---

In defense of the public interest, and in deference to the global realm, lifelong adherence to social morality and environmental justice needs to be legally enforced (an oath?), pledged at the time of graduation by all graduands. Why ever not? We should never again employ, receive advice, or allow space in the workforce for any professional, in any discipline, unless and until they pledge to a morality that sustains environmentally lasting welfare and upholds socially worthy outcomes. Proven miscreants, leaders and quisling alike, must face and suffer exemplary incarceration.[137]

'*Quantitatively Less*' in the future means '*Qualitatively More*'; in fact everything worthwhile satisfying and enjoyable in terms of transitioning to and sustaining a wholesome, lasting, and of necessity more modest, lifestyle. Indeed, within the OECD 'minority' context, a halving of personal disposable income would still vastly exceed the personal average within the 'majority' poor nations.

Resolve is required to get 'back' to locally balanced living which gives up on debt-driven growth blandishments, and the GDP fixation. This reasoning is advanced by Wilkinson and Pickett's empirical research (*The Spirit Level*) supporting the contention that misery profiles most within societies of notably unequal-wealth distribution; and conversely that misery profiles least in societies of more-equal wealth distribution.

Also telling is research by Derek Bok (*The Politics of Happiness*) establishing that high-end wealth and happiness do not correlate. In terms of lifestyle quality these three author's establish something most of us believe we know intuitively, that modest-wealth societies are relatively more balanced and happier societies. A transparent and stable money system providing a non-speculative service to society is the key to a sustainable future.[138]

## POPULATION STABILITY

Influence and empower birth-cohort women, with partner support, to 'stop at two' children, and support others who 'stop at one' or 'none'. The population concern and the overpopulation limit are matters most international authorities and research agencies choose officially to neither discuss or profile, being tacitly unmentionable. The reasons range from avoiding political embarrassment, to appeasing those factions promoting an anti-intervention agenda. It is also, of course, awkward to advance reproductive limitation in poor communities in the face of the need to maintain life support in old age. Also pro-life defence is a 'to die for matter' to some, regardless of the lower quality of life for every extra survivor-consumer.

If, as projected by WHO and UNESCO, the current seven billion of population expands to ten billion, then additional to an inevitably mangled economic outcome there will emerge an even larger population living reduced lives within degraded habitats. Growth in population is clearly an impediment to the future quality-of-life on a finite planet. Simply put, here on earth we cannot be all wealthy; yet we could mostly be healthy; and more than sufficiently well off with a lesser carbon consuming population (of around six billion?).[139]

---

**The advocacy here is to encourage all manner of debate and examine every angle in the pursuit of policies for human fertility decline; explicitly, a matter of getting birth rates below death rates.**

---

The principal environmental problem is wealth expansion and wealth leverage; the main inferential problem is the expanding consumer mass. In other words the journey to an environmentally stable future for humankind is co-reliant on pulling back wealth-inflation and population inflation. One social science label for this reduction of both consumption levels and fertility rates is distributive justice; a climate balance arrived at by halting outrageous expansion in the financial sector, coupled to gradually and sensitively managing fertility decline.

The lessening of population expansion (fertility decline) through a demographic transition has 'hard' and 'soft' attributes. The clear preference is for willing, educated and enlightened people to produce fewer doubly-loved children.

## PROMOTING FERTILITY DECLINE

- Establish reproductive health services
- Improve child survival to reduce fear of isolation in old age
- Empower and protect women (equality)
- Legalise elective birth termination
- Educate girls: including reproductive health education
- Statutory security net for the old aged
- Leadership: political, community, religious

Two more subtle items in the Sachs listing are Green Revolution Benefits and the Consequences of Urbanisation.

This stricture is neither draconian nor immoral. It is a matter of how to influence and empower people, particularly the under-educated poor, to ease family size, rationally, downward. The diagram on page 109 lists ways and means for inducing a reduction of fertility rates; components examined authoritatively in full by Jeffrey Sachs in *Common Wealth* (2008).

Island-nation contexts clarify the principles and illustrate some of the benefits. For densely populated poor island populations (Bali and Sri Lanka) an eventual halving of human numbers would lead directly to a doubled living space, and provide a more stable and supportive habitat. For also densely populated but contrastingly wealthy island entities (Singapore and Hawaii) protection of the environment through emissions reduction requires both a smaller population and a co-lessening of fiscal leverage, consumption, and discard, which is altogether more challenging.

There is also an inevitable and doom laden urban spectre to consider in terms of population. It arises because, as noted earlier, for the first time in human history urban places (which are almost entirely resource consuming and waste discarding) are where people are now piled-up. With little in the way of food-producing urban space, continuing to reside less mobilised in cities throws up new challenges and new 'living lightly locally' initiatives.

Extended families and co-operative groupings can be fed, clothed, schooled, entertained and protected on a community basis, achieving a much smaller per-person carbon footprint. Throughout the former USSR much specialised niche employment has evaporated, building maintenance has failed, and in many places support utilities and services have ground to a halt. Survival being the mother of invention, people have moved out of binary family apartments into knocked-together co-operative housing as a rational alternative.

---

Another community-support option, also emerging in nations of wealth, is the co-housing arrangement. Then there is the broader inclusionary Mixed Urban Development (MUDs) concept for co-mingling compatible home-places, service facilities, work places, schools and shopping in order to limit the wasteful movement of people and goods.[140]

---

For some, exodus out and away from the city will be an option;

encouraging people to couple rustic skills with electronic know-how, and use the food, fibre, construction and energy resources locally available. Others, in generously spaced and rain-watered lower-density suburbia, should be able to modify their home-and-plot holdings; retrofitting to solar power, water harvesting, waste recycling, and vegetable gardening

Nations which perpetuate belief in the current print-and-pump monetary system as teflon coated relative to nature will suffer their own, their community and their children's wrath. And the converse? Nation-states running stable monetary systems, all the while lessening carbon consumption and nitrate manufacture while evolving sustainable local communities for a lesser population, will reap the appreciation of their children's children.

---

**Summarising climate stability policy in terms of economic and demographic parameters: disengage, regulate, reverse and adapt away from the environmental degradation brought-on through inflationary money supply and a mass consuming over-population.**

---

The next three 'Action & Adaptation' chapters — Resource Conservancy, Goals, Targets — feature the adoption of pragmatic resource abstraction and waste discard protocols. Prevention and adaptation is the key.

Our challenge is to be fair to those who inherit, improving the lifestyle quality of both the subsequently less-rich and the subsequently less-poor. It is a challenge requiring wisdom and intelligence; which means that it is also hugely fulfilling as well as broadly sustainable.

## FURTHER READING

ECONOMICS:
Brown, *Eco-Economy*.
Daly and Cobb, *For The Common Good*.
Benes and Kumhof (IMF), *Chicago Plan Revisited*.

DEMOGRAPHICS:
Ehrlich, *The Population Bomb*.
Loveridge, *Vanishing Face of Gaia*.
McKibben, *Maybe One*.
Sachs, *Common Wealth*.

# 12

# RESOURCE CONSERVANCY
# with WASTE CONSERVANCY

---

**Contraction and Convergence | Carbon Taxation | Carbon Sinks | Small Scale Project Guidelines | Ecologically Sustainable Guidelines | Technologically Appropriate Guidelines | Input-Output Modelling**

---

Resource Conservancy policies and practices and Waste Conservancy policies and practices are integral to environmental defence; with improvements from both, and efficiencies to each, amounting to a sum social benefit for climate stabilisation.

Reviewing 'Economics and Demographics' in the previous chapter established that it is evasive to consider one set of options separated from the other. These policies and practices have synergy. The binding feature (for this chapter) is reduction of carbon resource uptake in combination with parallel efforts to improve waste reuse and recycling; and to ensure that a stable population pursues this goal (next chapter) for the common good.

Globally, mineral-oil ground-water and clean-air are all getting tapped-out; defining the need for efficiencies in the use and uptake of these resources, and the blocking of inefficiencies associated with the disposal of damaging residuals. A mistake with climate control advocacy has been the separated expectations of resource conservancy and waste conservancy. This has allowed the wealthy to make a show of offsetting their consumer wickedness. For example: Importing Perrier water from the French Alps to consumers in The Antipodes, who may then pay stylish attention to discarding the cute empties; possibly flaunting their faddish mobility selection by running to the bottle-bank in a hybrid car!

Decreasing resource consumption, along with less waste and improved waste reuse, can be tracked through resource conservancy and waste conservancy audits. The benefit of such an input-with-output system of assessment (elaborated toward the end of this chapter) is that it can provide progressive dual-tracking, calibrating resource conservancy and waste conservancy within the same construct, the soft pathways into the future.

---

### Appendix C — Exemplars From Oceania IV
### Papua New Guinea: Everlastingly Sustainable

---

Another way to illustrate the utility of resource conservation with waste conservation is to range living heavily in a 'devious way', against living lightly in a 'clever way'. It is primarily the inefficient squandering of finite mineral carbon capital (oil, gas, coal, shale) to produce human comfort, satiation and mobility that must be addressed. The problem here is that suggestions for reducing the carbon footprint, notably through emissions trading schemes, are specious to most ordinary mortals, and welcomed with get-rich glee by speculating investors.[141]

---

**Efficiency is the keyword; contraction-and-convergence the phrase; more fully 'contraction of inefficiencies and convergence of efficiencies'.**

---

Think now in terms of reducing fossil fuel use: improving home heating and cooling, constraining people and goods movement, improving the effectiveness of transportation and communication, increasing the degree of vegetarianism, localising agricultural uptake and food processing, and seeking-out energy savings in the workplace schoolplace and marketplace. Think also in terms of less waste generation and resource recovery from trashed materials. In jurisdictions of wealth the opportunities for these outcomes, mainly as carbon savings, are realistic and considerable. The point and purpose of these gains is that they are operationally effective and environmentally beneficial.[142]

Staying with 'contraction and convergence': contemplate food available for on-the-lips consumption, again mainly in the context of wealth. Modern agriculture (and feral fishing) is beset by mechanisation and trundling and trawling every which-way; often delivering food a hemisphere separated from the source of harvest or capture. Also, in the case of agriculture, after being cultivated, fertilised (with nitrates and phosphates) and then harvested, comes processing chilling transporting display purchase preparation cooking and waste disposal — all activities adding to the fossil energy account. This provides an opportunity to emphasise near-vegetarian and tax-on-meat options and to locally produce food for the home kitchen: achieving energy efficiency on account of a much lesser fossil fuel input, leading in turn to a lowering of $CO_2$ nitrous oxide and toxic ozone emissions.[143]

Contraction with convergence is a potent policy-combo for stabilising climate, with beneficial environmental consequences for both the impoverished and wealthy. Of course it is the smaller populations of greater wealth who must deliver results quickly, with the larger mass of poorer people sustaining a lesser contribution in aggregate over time. As noted earlier there can be technological efficiencies across the board; in industry and agriculture, in the home-place, office-place, school-place and shopping-place, and in transportation and communications. Here it is realistic to note that, of course, we do not all have access to the same options.

In times of fiscal buoyancy it is pragmatic to direct money to environmentally approved projects, and to give incentives to avert environmentally degrading processes. But the revenues gained and the monies paid out become environmentally irrelevant *unless* the take-up systems are squeaky clean and the money is applied specifically to achieve demonstratively effective environmental cleanup.[144]


Emissions Trading Schemes rely on complex easily evaded and readily manipulated processes for reducing greenhouse gas emissions. As a procedure ETS has so far failed to reduce levels of heat-trapping carbon and methane gas globally, and the same outcome holds for every participating country considered separately; a problem here being that dense-energy fossil fuel use remains so reliable, available, portable and affordable.

In a transparent and strictly honest cost-and-profit world a mandatory ETS would work by allowing carbon emission taxing (the cap) to be run alongside carbon emission brokerage (the trade). A problem is, that operated in this way, the ETS arrangement sanctions, indeed encourages, emissions from the likes of a coal fired industry in a wealthy economy, offset through alleviation payments made for allocations in a poorer and less carbon emitting economy.

There is also the matter of inequitable comparability. As examples: some schemes exempt agriculture others don't, most exempt aviation a few don't, others credit plantation forestry as carbon sinks some don't, others exempt coal-fired power generation some don't, others exclude transportation fuels some don't. None control or tax methane gas from ruminants. In all these contexts the efficacy of ETS is underwhelming.

What we know from human aptitudes for fiscal trickery is that no matter how well policed and enforced, evaders will work out ways to sidestep the cap (tax avoidance), and also work out ways to bend the strictures to their financial advantage (rules evasion).

---

**In order to work administratively there has to be rock solid agreements on enforceable standards country-by-country involving all jurisdictions; backed-up with a foolproof and accurate MRV (Measurement Reporting Verification).**[145]

---

Variable progressive carbon taxation prior to carbon uptake and consumption is politically unpopular; yet this approach is more direct, purposeful and enforceable than the operation of emissions trading. The important issue is that trades involving the portability of greenhouse gas allowances are eventually a disbenefit to the global biosphere. Engaging the already discredited and alien-to-nature fiscal mechanism, via an ETS format, to correct environmental failings is flawed; it amounts to thinking from within, rather than outside 'the banker's box'.

Emissions Trading Schemes encourage systemic avoidance and systematic non-compliance. When confronted with a fiscal downturn the procedure is also vulnerable to political vacillation. It is strategically disadvantageous and clearly disingenuous for any nation, unilaterally, to go out on a limb with a stand-alone ETS format. To do so simply encourages emissions-intensive industries to move offshore into the arms of jurisdictions with an unused ETS surplus, costing onshore jobs.

A first problem (highlighted in the previous chapter) is that emissions trading encourages pork-barrel favouritism, with little regard for the environmental consequences. A second problem is that encouraging these policies fosters the emergence of an evasive black economy, getting away with harmful carbon gas and other discards. A third problem is that holding the line on emissions trading goes out the legislative chamber window during times of economic recession.[146]

---

The primary drive is for effective carbon gas reduction and stabilisation; the most direct route being progressive carbon taxation in association with incentives for lifestyle contraction-and-convergence.

---

Additional to carbon capping and carbon taxing is the practical need for reforestation, the Permanent Forest Sinks Initiative for soaking-up excess carbon dioxide; along with the installation of some yet-to-be-proven chemo-mechanical carbon-gas capture devices (refer back to endnote 116).

## SPECULATIVE CARBON TRADING

Hedging the future global price of earth-bound mineral carbons, and corn not yet sown, locks in inflated future prices to the consumer; profiting traders who take a punt on the unlikely event of being caught short in an upwardly spiralling protection racket. The social point of allowing this to go on escapes me, always has.

Also unseemly is gaming the future price of carbon emissions by putting industrial carbon gas 'credits' into a supply and demand situation for brokers to trade. My rejoinder to this being that if they are out there as fairly adjudged amounts, where arises the carbon reducing benefit to the global biosphere from allowing them to be traded? This reasoning supports the preference for taxing to support conservancy at the point of carbon purchase, coupled to strict emission penalties and cleanup practices.

The greater need remains: to confront the harsh reality of increasing oil and brown coal consumption, thawing Arctic Circle tundra, ice-cap meltdown, wildfire conflagrations, deforestation of the Amazon Congo and South Asian rainforests, and the significant contingent problems arising from industrial nitrogen fixing and phosphate mining.[147]

These under-audited situations highlight the utility of an input-output array for understanding both the overall situation and its parts. Input-output, coupled to localised surveillance, can profile the environmental cost of manufacturing, agriculture and trade; and delineate policies for restricting pointless goods exchange, and the delivery of out-of-season food, international holidaying, cross-region commuting, rainforest clearance and the like. Waste avoidance, waste recycling, and waste reduction benefits can be also cranked through such an array. Efficiencies within the 'resource conservancy with waste conservancy' paradigm involve an alignment of astute protocols with fit-for-purpose delivery; which is of course much more easily stated, and put about as policy, than delivered.

---

**Actuate— Small Scale Project Guidelines, Ecologically Sustainable Guidelines, and Technologically Appropriate Guidelines.**

---

What follows is the braiding-out of these rubrics; emanations from the political drivers, demographic urgings, and economic essentials established in the three previous chapters.

**Small Scale Project Guidelines—**

Incorporate the axiom 'Think Global: Act Local' keeping in view that the practical outcome being most sought is an overall reduction of carbon gas emissions. To the sharp end of that reasoning: if people of wealth stopped moving about so much, reduced their consumption of accessories, eased-up on the movement of goods, rationed their uptake of services, and localised their work, schooling, vacationing and entertainment, then a lessening of emissions really would occur. In addition, smaller scaled and localised farming and manufacturing, and local power supply and waste disposal, along with a doubling of energy use efficiency, produces effective cuts in the consumption of mineral carbons.

Localisation of government (community subsidiarity) fits the small-scale scenario, along with local administrative help with the deployment of local skills and local provisioning; all accompanied by local waste absorption recycling and reuse. This emphasis on closer-to-home governance is, instrumentally, about reducing the movement of people and goods, and thereby lowering the overall level of mineral energy use.[148]

**Ecologically Sustainable Guidelines—**

We can add subtlety to the Small Scale Guidelines *via* efficiency; for example with the energy saved (uptake avoided) in ecologically aware agriculture (agro-ecology), manufacture, environmentally conservative operational practices, and the recycling of consumer durables and other waste. There is also the matter of mineral carbon calories 'saved' as a result of every food calorie produced and consumed locally.[149]

There are myriad ways to reconfigure and reduce the urban levels of household and personal footprinting. These include — as an extension from the Small Scale emphasis — mixed urban land use patterns, increasing the occupancy rate of buildings, improving the energy efficiency of construction, building insulation, kitchen gardening, stormwater capture, wastewater reuse, onsite energy sourcing, recycling, and the composting of organic discards. There is also the matter of an altered mobility, walking or cycling along with entertainment in or nearby to the home.

With farming there is, in addition to the reduction of atmospheric $CO_2$ concentration, the matter of halting nitrate build-up in the hydrosphere, permanent tree planting, the retention and conservation of ground water, the selective use of proven biological controls, and the localisation of produce sales.

Another item to consider is GreenTick information about manufactured products at the point of sale: data indicating the specifics of manufacture, carbon energy used in fabrication, the rate of carbon energy uptake and efficiency when in use, and the use of protection tariffs. Throughout, the quest is to maintain an ecologically sustainable efficacy — 'contraction and convergence' no less.[150]

**Technologically Appropriate Guidelines—**

These reinforce the previous 'scale' and 'ecological' guidelines, endorsing the celebration of many modern achievements. To the fore is the cellular telephone the internet service and the World Wide Web; ubiquitous and beneficial all-connecting utilities adopted globally by rich and poor alike, bringing news information and opinion from the rest of the outside world to every desktop cellphone and laptop. Then there is the wizardry and the option of solar, hydro and wind energy capture as carbon-free alternatives for powering-up and heating-cooling the homeplace workplace and schoolplace; along with the myriad technical, material, thorium nuclear and electronic innovations now permeating construction, manufacture, home living, industry and farming.

---

Relative to all three **Scale, Ecological,** and **Technological** efficiencies: **Bypass, Reject** and **Circumvent** environmentally adverse inefficiencies.

---

The tools available for managing healthy survival have their origin with the Precautionary Principle (identifying avoidable harm), which touches base with the ancient Pareto Rule (compensating for collateral injury to third parties). In our daily round we individually (mostly subconsciously) apply the Pareto Rule and the Precautionary Principle. We also have a pretty good understanding of how public-cost and public-benefit works out. For example: benefits (including emissions avoided and-or sequestered), and disbenefits (including emissions discharged free-of-cost to the consumer).

Input-output modeling in sub-national regions of concern is more purposeful than benefit-cost assessment; being time-lineal (past-present-future) and capable of incorporating non-fiscal quanta. Listed down the side of an input-output tabulation is data on population, investments and resources (as with workforce, monetary, carbon and nitrogen inputs) — basics that can be further broken down into farming, mining, industrial, and commercial activities. The across columns are dominated by the outputs (carbon emissions, nitrate runoff, entropic breakdown, wastes and losses, costs and benefits).

Input-output spreadsheet modellers are able to ascribe, with ever-improving assurance, climatic economic and demographic outcomes. Items like $CO_2$ ppm and nitrate concentrations, loss and profit, resource uptake and waste discard, changes to the population pyramid, and of course climate change can be shown; all from improved and refined fiscal, climatic, population, production, consumption and discard data. By this means the input-output process can influence the direction of demographic and economic policy in addition to profiling biospheric burdening. The significant point about an input-output spreadsheet is the interpretation obtained from past and present economic, resource and demographic variables, inference giving way to future actuality.[151]

Economic, population, consumption, and discard parameters can be expressed in an input-output construct for a region, project or community. From this it can be appreciated that the attraction of such an array is the way it provides understanding of the interplay between monetary, demographic, climatic and consumer-discard elements. Input-output spreadsheets show the consequences, adverse or favourable, in an explicit manner: more or less carbon dioxide and nitrates, less or more temperature change, stabilised or reduced biodiversity, larger or smaller ecological footprint, fiscal profit or fiscal loss. An input-output spreadsheet is of course challenging to set-up, and demanding and costly to maintain. Look on such a facility as an accurate and useful tool for fashioning (advising on) stabilising an environmentally favourable outcome.[152]

Input-output displays which profile the state of environmental health and the climatic effects of atmospheric and water-body waste discard can also highlight the need for prevention and adaptation — along the historical lines of the 1980s CFC studies which led to compulsory correction. If we know, with scientific certainty and statistical accuracy, the causal trends of climate change, then we are fact-based to call the shots for adjusting away from the indicated adversity. Also, in these terms, we can understand limits and initiate constrictions and efficiencies.[153]

Lessened resource uptake and improved waste avoidance are practical contract-and-converge pathways, noticeably effective when adopted by the wealthy. For the relatively poor the emphasis is on those innovations which are also lifestyle enhancing.

---

**What we arrive at, for both rich and poor alike, is that the current 13% inefficiency of fossil carbon energy use (with 87% 'lost') propagates a mantra — double the efficiency of carbon use and halve the rate of carbon uptake.**

---

Doubling, then moving on to treble the efficiency of oil gas and coal use from 13 to 26 to 39% and beyond presents science with a straightforward and relatively easy task. A knock-on problem being that fossil carbons sold at an unchanged price would induce consumer's to enjoy the benefits, to them, of an unaltered uptake. Doubling-and-doubling the tax could be an answer; leaving open the matter of how the tax is spent to environmental advantage.

Other pragmatic challenges for halving fossil carbon uptake include the lesser movement of people, the lesser movement of food, the lesser movement of water, the localised provision of construction and manufacturing materials, and localised solid and liquid-state waste disposal.

It is a matter of personal and community accomplishment to practice resource conservancy and environmental defence, to upkeep a waste recycling regimen, and to live lives guided by sustainable principles. In this an important ingredient is 'sense of place' our attachment to family and locale, and the thrill of community and belonging.[154]

---

The pursuit of 'resource conservancy with waste conservancy' instils moral uplift as well as pragmatic achievement, a through-line to setting the mainly community Goals and hitting the fundamentally personal Targets — the next two chapters.

---

### FURTHER READING

Schumacher, *Small is Beautiful.*
Hardin, *Living Within Limits.*
Arendt, *Growing Greener.*
Barnett, *Blueprint for Action.*
Bernhart, *Deeper Shade of Green.*
Owens, *Energy Planning and Urban Form.*

# 13

# GOALS

---

**Stabilise Money Supply | More Than Halve Fossil Fuel Uptake | More Than Double Carbon Use Efficiency | Quarter Nitrate Manufacture and Use | Negotiate Birth-Rates Below Death-Rates | National Goals | Community Objectives | Household Responsibilities**

---

This chapter moves on to can-do matters. In so doing keep in view the essentially social purpose that underlines and vindicates climate stabilisation; worthy, wholesome and balanced human continuity.

We know of the need to at least halve reactive nitrate pollution and greenhouse gas emission, yet practice has proved contra-effective. The main social marker put down earlier is 'why' these practical objectives should be fashioned into an ethical goal? The retort: for the future of our children's children. But here there is a problem in that the economically advantaged of the so-called First World have yet to acknowledge and respect biophysical limits of resilience, let alone set about achieving goals for reduced carbon and nitrate discard for the Majority World.

Embedded though I am in a lifestyle of modest wellbeing, I am also reflective of poverty, about which I have an understanding from a range of research and development work in poor jurisdictions.

Environmental good work in the wealth context genuflects to sustainability, mostly with an unintended shortfall. In my low density home town, for example, behemoth-trucks belching carbon fumes call by each week to pick up the recyclables — that is trucks plural, one for paper, the other for cans and bottles — any benefit being doubly cancelled through the delivery of the morning newspaper by a guy in a 3.6 litre SUV.

Environmental good is of course achieved in part by some wealthy and moderately wealthy enthusiasts; people who make an effort to score a degree of personal resource conservancy, personal waste conservancy, and a personally reduced carbon footprint. They are the eco-enthusiasts of wealthy societies; yet compared to the inbuilt unconscious conservancy of the poor in the Third World, the sum of their attainment is paltry. The point here is not to belittle the effort made by the well-to-do, but to understand these efforts for what they so far achieve, an insignificant contribution (in effect a non-contribution) to the planetary task of reducing greenhouse gas emissions overall.

Eco-advice emanates from the largely Transatlantic and Australasian environmental press. This literature is enthusiastic and heartfelt; but it can be trolled in vain for mentions of wealth as a problem (beyond ethical investment guidance for the eco-squeamish); and there is scant attention to population overload (beyond an occasional 'stop at two'). The volume of published output listed in the References is testimony to an earnest concern to make a difference on the basis that the sum of all the actions and advice disseminated will reduce the adverse effects of global warming. Surely? But, actually, not yet by much![155]

A favourite among these writings is *World Changing* (edited by Alex Steffen along with a host of contributors); with some Third World insights and myriad suggestions on how to cope with living in an environmentally appreciative way involving green business, green product and green certification.

My copy of *World Changing* lists two hundred compatibly focused wealth-society publications in its Selected Bibliography. Other 'lighter green' favourites *The Lazy Environmentalist* (Josh Dorfman), and *Green Chic* (Christie Matheson) are more frivolous, verging toward outrageous. These endeavours are laced with humour, which include thoughts on such environmental concerns as 'pets', 'travel', 'wardrobe' and 'decor'.

*Solutions to Climate Change* produced by Guy Duancey and Patrick Mazza, is focused Stateside and Canada (Organise a Car-free Sunday!) with a prodigious listing of web sites. And for out-there techno-wizardry consult Chris Goodall's *Ten Technologies to Save the Planet*.[156]

---

### Appendix C — Exemplars From Oceania V
### Cook Islands: Technological Sufficiency

---

The one-fifth of the world's population that is well off (conspicuously the New World Settler Societies, Europe, Japan and the 'petro-dictator' nations) do not want to relinquish their opulence. The other four-fifths relatively quite poor (dominantly China, India, Pakistan, Indonesia, most of Africa and much of Latin America) want their material prospects to improve.[157]

Those of wealth if asked to downsize say 'no'; those in poverty if asked to moderate their upsizing also say 'no'. Impasse: for neither the rich nor the poor can find a way to agree about how to meet environmental goals through monetary adjustment and a reduction in their uptake of fossil energy and consumer 'goods'. Clearly the rich-poor divergence has to be reconciled; differential constraints have to be worked out for the wealthy and the poor.

---

**In relation to consumption the necessary 'dematerialisation', in general terms, is for the wealthy to at least halve their overall fossil fuel uptake and reactive nitrate use; and for the fossil fuel consumption and nitrate uptake of the poor to be constrained.**

---

In the context of economic reasoning the primary objective is money supply control and tax haven closure (see Economic Stability diagram on page 105). In the context of population numbers the main goal is guided reproductive stability (see Fertility Decline diagram on page 109).

At this point it is relevant to return to a through-line, identified earlier as pragmatic 'efficiencies'. Four run-together contexts can be tagged.

1. Practice day-by-day lifestyle conservancy, essentially a matter of striving to live mainly off renewable resources.

2. Operate small projects on a local scale, at a slower pace, working with ecological forces and within biospheric givens.

3. Engage and celebrate the use of those gizmos which are benign and useful (solar wind and water energy capture devices, electronic information sharing, future-proofed GM plants, and non-toxic chemicals and pharmaceuticals).

4. Reduce, adjust, localise and downsize consumer (notably industrial) waste discard, which includes the efficient rationalisation of social and goods movement.

OVERARCHING: Manage-Up (Increase) the Efficiencies, and Manage-Down (Reduce) the Inefficiencies.

My working life has been devoted to planning. Producing, delivering and adjusting plans: development plans, conservancy plans, regional plans, town plans, tourism plans and one intentionally deliberate, fully consulted and detailed village plan. Work of this kind nourishes a proactive, problem-solving and potential-realising fieldwork style. And as this is the toolbox I know it proscribes my approach to this writing.[158]

Planning processes are complex and contingent, but far from arcane. Simply put; planning involves focused forethought before initiating rewarding action. Those who have made a few investor millions can be styled wealth planners, those who have scored some property deals can be styled speculator planners, those who have tickled the stock exchange to make a wadge can be styled portfolio planners; we are all planners by intent and experience. The process is well understood. I have less experience, but some, of de-planning, un-planning, planned retreat and planned withdrawal; and that approach and technique is in part the key to attaining future economic stability, fertility balance, reduced carbon uptake and reduced nitrate use.[159]

As noted in Chapter 10 a GlobalGood service would bring to prominence all that is known about the factors influencing and threatening climatic and biophysical change (principally the regressive economic-demographic synergy) and deliver guidance on prognosis, policy and practice. And what better place for putting out day-by-day reportage and listings on climatic, atmospheric, biosphere, sea level, economic and demographic 'spot positions' than the already 'fit for purpose' desktop screen and daily press.

To that end it would be useful to provide a quarter-by-quarter and year-on-year output on economic stabilisation, rates of population fertility change, circumstances of biophysical change, and case notes on environmental justice. This information could be published as a timeline tracking of factors that contribute to both sustaining and damaging the biosphere.

Envisage caring, sharing and informing through the outreach of virtually costless, electronic information dispersal. Envisage adopting more locally focused and interdependent lifestyles within proscribed resource uptake and waste discard limits. Envisage a reduction of unnecessary external trade. Envisage achieving a sensitively programmed no-growth population stasis. And also envisage living lightly locally with a reduced movement of goods and people — all the above tracked and checked via a GlobalGood monitoring protocol.[160]

Being **Clever and Wise**, and **Sustainable and Soft**, and **Positively Proactive** are policy and practice drivers to lace into the 'efficiency-up inefficiency-down' and the 'contract-and-converge' rubrics detailed in the previous chapter.

**Cleverness and Wisdom** is a combination with political as well as practical characteristics; for example a consumer durable can only be acceptable in socio-environmental terms if it is manufactured inoffensively, works efficiently, can be operated and maintained easily, and has a recyclable end-use. Hold in view that the earlier reviews of pan-political drivers (Chapter 9) called for the connection of economic to environmental values.

The other significant call was to divert from unnecessary consumption (a big ask for the governments and populations of North America, Australasia, Europe, and the wealthier oil producers). In all this the 'elephant in the room' is China; with burgeoning consumption, dubious but improving labour and pollution standards; yet to the fore and forging ahead with the manufacture of solar hot water systems, windmill technology and permanent forest replanting.[161]

Returning to the tracking of outcomes. One mechanism styled as Kyoto2 (attributed to Oliver Tickell 2008) would be a quota-based uptake of finite mineral carbon resources, governed by objective (essentially non-market) systems for regulating and taxing supply. With such an approach it would be essential to ensure that licensing and taxation revenues are used to fund technological adaptation and mitigation measures.[162]

An accessory to consumption control, as noted previously, is Production and Origin information displayed on packaged goods at the place of purchase. Data of this kind is useful when accompanied by data on the achievable discard gains when consumer items are dumped. That knowledge, linked into Input-Output displays, produces a library of information to be analysed and put out in GlobalGood bulletins.[163]

Food supply, and manufacturing and construction processes, can be adapted to a lower carbon uptake and a reduced nitrate input, reducing atmospheric carbon dioxide excess and hydrospheric nitrate over-enrichment. In alignment are the alternatives for producing energy from freeflow sources — photovoltaic power systems, hydro power plants, wind turbine facilities and geo-thermal installations.[164]

Recycling household rubbish and abating industrial waste (the 76% discard factor) is important. It is environmentally astute to initiate waste capture at the point of emission or rejection; and to recover rehabilitate and reuse waste discards, known to some as a cradle-to-cradle process. Control is achieved by enforcement of the Polluter Pays Principle, and the requirement that scrupulous penalty and compensation payments are applied to cleanup.

The engagement of wisdom (which amounts to applied cleverness) is a complex matter to which this insight is a taster; confirming what we realise yet have lacked the personal initiative, and official guidance, to achieve.

**Sustainability and Softness** is about continuity of socio-environmental balance indefinitely — probably best put as 'foreseeably'.

Applying the sustainable ideal is guided by the characteristics of the three categories of resource involved. In the case of *Free Flow Resources* (wind energy, tidal and river-water hydro, direct solar) these stream to us from continuous supply sources and are available for capture and exploitation at will in a socially responsible way. The principles of sustainability come fully into play in relation to the uptake of *Finite Resources* (mainly fossil carbons and atmospheric nitrogen) where control of use, restraint, regulation of uptake, and taxation is vital to maintain a continued yet much diminished and environmentally less damaging supply. Then there are the ever-replenished *Living Resources* of the biosphere (soils, oceans, atmosphere, aquifers, flora and fauna) over which the practice of astute husbandry is required to maintain a healthy and continuing balance.

Gentleness suffuses sustainability, which I characterise as softness. The 'soft pathways' (first identified in Chapter 6) respect the limits of biospheric resilience over profit, favours localised small projects over mega developments, operates in a biologically harmonious way, sustains wholesome life through biospheric regeneration, reduces dependence on bottled sunshine (fossil fuels) in preference to solar energy capture, retains and fixes nitrogen naturally in agriculture, and encourages a light-and-local lifestyle.

An expanding population can nullify technical gains in energy use efficiency. Factually, in relation to population, an unknown unborn unbeing does no consuming and has no potential to produce additional offspring. Furthermore, when communities remain small and exhibit self-sufficiency, the global capacity for living within resilient parameters improves. Here is where eco-design cuts in to use freeflow resources, environmentally benign construction, mobility alternatives to petrol driven vehicles, and homeplace climate controls.[165]

Much clever technology is available; electronic monitors and data sharing devices, photovoltaic's, wind and hydro generators, chemical engineering, improved plant breeding, and the edible treescape. All support a more sustainable habitat emanating from fiscal certainty, the slower exploitation of finite resources, a stable population, and the improved husbandry of renewable resources.

**Positive Proactiveness** involves delivery to the sustainable imperative. There are various ways to penetrate this topic, with the focus now on principles, followed-up with more detail in the next 'Target' chapter. The metaphorical millstones grinding away through these pages induce improved climate certainty *via* a better than halving of mineral carbon extraction and atmospheric nitrogen fixing, along with a better-than-doubled energy use efficiency. The overarching necessity is to 'live lightly locally' and to kick the carbon energy habit out-of-play in most aspects of our lives.

Much of the focus is urban because the concentrated emitters, and those of wealth, are urban based. Powered mobility can be reduced by capacity (smaller sized vehicles and increased use of public transportation), reduced intensity (fewer trips per day), reduced array (contracted geographical range); and avoided altogether by replacing motorised trips with cycling, walking and electronic communication.

In order to further reduce automobile dependence the principle correction to urban community arrangements involves the interrelation of home-places, work-places, shop-places and school-places so that they are within cycling and walking proximity each to the other. This is known as the MUD concept (mixed use development: permeable, walkable-cyclable, largely self governing and self-policing) avoiding the cross-town frenzy inherent in separated-use zoning.

The MUD concept approximates to some of the layout and micro-design principles of 'new urbanism'; a largely untested extension being the eco-village ideal, setting out to achieve all manner of pedestrianising, insulating, energy conserving and recycling objectives within a largely self-sufficient cluster. There is also the engagement of high conservancy standards for the design and fitting-out of buildings and services.[166]

The major challenge, particularly in the New World settler societies is to retrofit existing drive-only suburbia. Suburbs comprising multiple-generation households on larger (more than quarter acre) plots are potentially viable in a variety of self-sufficient ways; involving reduced reliance on the use of automotive transportation, reduction of inbound food water and energy supplies, the onsite absorption of wastes, and kitchen gardening. These are principles considered for the initially British (Totnes) Transition Towns, an urban-based contract-and-converge initiative now spreading to other countries along with the Italian Cittaslow movement.[167]

In the matter of 'green building' design the objective is ZED (zero energy development, which is approximately the same as ecologically sustainable development ESD, and the North American LEED 'leadership in energy and environmental design' programme). Here the aim is to adapt and design residential buildings to capture maximum energy from the sun, collect rainwater, recycle waste, and be insulated and ventilated against extremes of hot and cold. The outcome intended is a massive reduction of each household's imported energy, in line with a reduction of offsite support for onsite human sustenance and comfort.[168]

Buildings erected in the past, in moderate to wealthy societies, exhibit several environmental failings including unused space indoor and outdoor, and an ongoing energy demand for climate control. This opens up rehabilitation options yielding massive permanent energy savings. Onsite energy production and improved energy-use efficiency is partially practical. Clever non-electric refrigeration and air-conditioning is now available, along with improved heat-exchange (pump) technology. Compact fluorescent bulb use is becoming standard, fuller building insulation is simple direct and practical, appliance efficiencies and standby capabilities are improving, waste recycling makes simple common sense, and there is the potential of green surfacing and kitchen gardening.

Farming the rural landscape for food and fibre opens up parallel possibilities for conservancy. Here the call again is for less mechanisation, lowered applications of manufactured off-farm nitrate and phosphate fertilisers and chemicals, greater emphasis on biological controls and the use of future-proofed GM cropping, less pastoral and more arable production, a local marketing focus, increased planting of food fruiting trees with inter-cropping, much reduced aquifer-fed irrigation, and greater rain-fed reliance and water retention; all coupled to superior natural source energy capture, waste avoidance and recycling.

In terms of both the urban and rural contexts, in locations of both wealth and poverty, it is the fossil fuelled energy habit that has urgently to be curtailed through coercion, regulation, enforcement and penalty. An aspirational plea for a reduction of fossil fuel consumption ('no more than two more degrees of future temperature rise') was endorsed 'in principle' at the United Nations 2009 Copenhagen 'Conference of Parties'; but only after turbulent non-binding negotiation by the major polar-opposite participants, China and the United States — pleading 'exceptionalism'. Then, from participants at the 2011 Conference of Parties, came the Durban Treaty pledging a timeline for the negotiated reduction of carbon gas emissions — a good start, for a process and outcome much more complicated than simply setting out to reduce carbon emissions. This modest progress was stalled at the 2012 Rio+20 gathering.[169]

The alternative energy-tech options are smaller, safer, longer lasting, reliable and, above all else, doable and sustainable. These supply-side facilities garner energy from where the water flows, the wind blows, the sun shines or steam puffs-up; the wind-driven, solar-powered, hydro-kinetic, heat-exchange and geothermal options. They all avoid or reduce reliance on hydrocarbons. Nations do not of course all have access to all these renewable options.[170]

Transmission losses are reduced when power is generated and consumed locally from wind farms and solar gardens. This simple precept achieves an energy 'saving', offset to some extent by energy 'expenditure' in the manufacture of installations, and the inability to supply continuous base-load power. The photovoltaic future looks hopeful with the manufacture of lengthy rollout panels (even PV paint) absorbing modest amounts of sequestered energy during manufacture.

Throughout the habitable parts of the planet, photovoltaics and windmills open up prospects for installing on-the-spot micro power plants, when and while the sun shines, water flows or the wind blows. Cheap and lasting battery storage is the key to moderating the peaks and troughs of photovoltaic and windmill supply.[171]

Relative to the human scale, hydro dam options are always big, sometimes enormous; and they usually prove problematic during construction and in operation. Aside from irrigation and flood control benefits, dams generate electrical energy. Hydro-dams abound in many OECD landscapes — Northern Scandinavia, North America, Australasia, with the behemoths in the poorer regions of Africa, India, China and the Amazon. Many are, or were, inefficient; and some are now unsafe. With refitted turbines and improved operational gains, coupled to the still distant prospect of near-superconductor line transmission, their future is assured.

Wind turbines installed at chosen locations as 'wind farms' in high latitudes, have an advantage over daytime-only 'solar gardens' and tide-dependent turbines on account of their twenty-four hour operation, (provided there is sufficient breeze). Again, the energy embedded into manufacture and installation on site is high — as is the less relevant fiscal cost. Most neighbours accept large turbines twirling noiselessly beyond hearing on distant upland. But there is always disquiet about the noise generated from closer-to turbines, and when attractive landscapes are occupied. There is also objection to smaller domestic turbines and their dangerous sounding clatter, their relative inefficiency, their danger to birds, and their high manufacturing and installation cost. Design and operating improvements, along with improvements to battery storage technology, continue.

---

**Adhere to a 'soft' techno-environmental credo: Smaller, localised, selectively hi-tech when appropriate; and also toxin free and waste abominating.**

---

Softer 'slow living' options have several characteristics. Here I revive a previous (2004) listing of 'Soft Pathways'.

These include:

- Smaller scale operating patterns (community based).
- Scaled down mobility (cycling and walking).
- Small sized projects (onsite energy and food production).
- Locally sourced construction materials.

Add to these the more useful of the high-tech utilities; cellular phones and other electronic information-sharing devices.

There are also some *'Ecological Pathways'*. These include:

- Greater and improved at-home food production.
- Fibre and nutrient composting.
- Biological pest control and fail-safe GM cropping.

Rounding-off the list is Waste Avoidance: Refuse-Reduce-Repair-Reuse-Recycle, along with Toxin Denial.

---

**SUMMARISING 'POSITIVE PROACTIVENESS':**
**Local Materials Sourcing, Local Food Production and Uptake,**
**Local Skills, Local Waste Management, Local Governance.**

---

There is also a mix-and-balance of impact considerations to weigh-up. These arise in the manufacture of consumer durables, along with the manufacture of pharmaceuticals and household chemicals; initially on account of the embodiment of mineral carbons in fabrication, ultimately because of the environmental impacts during use and disposal.

This returns to the case made earlier for eco-labeling packaged products; and depicting information about embedded energy on manufactured items as an extension of the packaged food labeling practice; giving purchaser's the opportunity to compare 'price tag' with 'carbon tag'.

This chapter has reached the stage where the essential Goals can be listed as **National: Community: Household**. These are set down prescriptively in what is, unavoidably, a lengthy litany; overarched by the mantra to 'increase efficiencies' and 'decrease inefficiencies'.

## NATIONAL GOALS

- Pan-Political Accord: Economic-Demographic-Environmental
- Sustain A Fiscal Stasis [Chapter 11 diagram]
- Seek A Less-Than-Replacement Population [Chapter 11 diagram]
- Sustain 'Environmental Justice': Rights, Defence, And Prosecution
- Endorse And Contribute To The Globalgood Enterprise
- Reduce External Dependency And Country-To-Country Trade
- Centrally Risk-Evaluate Large Projects: Use Call-In Powers
- Public Ownership-Management Of Mineral And Fresh Water Resources
- Meet Energy Needs From Solar, Wind, Water-Flow, Geothermal Sources
- Adopt A Green Procurement Policy For Public Services And Products
- Apply The Polluter-Pays-Principle At Source, To Users, And On Discard
- Quota-Regulate And Tax Carbon Supply And Uptake
- Assist Carbon-Reducing Energy Projects; Apply Eco-Harm Taxation
- Prevent, Penalise And Tax Carbon-Energy Profligacy
- Reward Self-Sufficiency, Carbon Neutrality, Import Substitution
- Localise The Provision Of Food, Fibre And Construction Materials
- Encourage Compatibly Mixed Urban Land Useage (MUDS)
- Mandate Embedded Energy Information On All Packaged Products
- Embrace The Subsidiarity Principle; Pass Governance Down
- Support Ongoing Long-Term Conservation And Sustainable Research
- Encourage Gender-Harmonised Management And Governance
- Minify Industrial, And Repudiate Toxic Waste Production
- Outlaw Fiscal Buccaneering: Discipline With Exemplary Rigour.[172]

## COMMUNITY OBJECTIVES

- Align And Fall-In With National Goals And Directives
- Control Non-Renewable, And Conserve Renewable Resource Use
- Practice Environmental Defence; Reduce Fossil Carbon Uptake
- Facilitate Sensitive Fertility-Decline Management
- Combine The Polluter-Pays-Principle With Cleanup Responsibilities
- Adopt Transport Demand Management (TDM) Strategies
- MUDS And TODS (Mixed Use And Transport Oriented Development)
- Support Multi-Generation Household Formation
- Enforce Restorative Penalties For Environmental Malpractice
- Enforce 'Green Building' Design And Retrofitting
- Facilitate Solar, Wind, Water-Flow And Geothermal Installations
- Waste: Reduce, Reuse, Recycle, Repair, Rehabilitate, Revitalise

- Reduce The Ruminant Animal Population
- Manufacturing: Source Local Materials, Develop Local Skills
- Encourage The Local Economy And Local Barter.

## HOUSEHOLD RESPONSIBILITIES

- Align And Fall-In With National Goals And Community Objectives
- Live Lightly Locally: Food, Energy, Construction, Work, Learning, Play
- Walk And Cycle: Use Public Transport
- Discards: Reuse, Reduce, Resist, Repair, Recycle, Retrofit
- Reproduction: Empower Women To Stop At Two, One, None
- Lifestyle: Move Ever Closer To A Vegetarian Diet
- Avoid Mineral Carbon Uptake And Use In Every Way Practical
- Use Mineral Carbons With Much Improved Efficiency
- Facilitate The Local Economy: Barter, Exchange, Co-operatives
- Obtain Energy From Solar, Wind, Moving Water, Geothermal Sources
- Intellectualise: Gain Personal Satisfaction From Service To Society
- Respect Nature, Dignify Labour, Honour Craft Skills
- Use Personal Wealth In An Environmentally Judicious Manner.

While the next chapter on Targets narrows down to the detailed and practical matter of specific actions, it is essential that the origin of these actions be recognised as coming from a greater exigency: the climate-stabilising challenge to fashion, adopt and apply international protocols. At legal base it is a matter of Conservancy Practice and Environmental Defence, adhering to the protocols of Environmental Justice.

### FURTHER READING

Christensen, *Innovators Dilemma.*
Steffen, *World Changing.*
Desai & Riddlestone, *BioRegional Solutions.*
Lewis & Gertsakis, *Design and Environment.*
Hopkins, *Transition Handbook.*
Langdon, *A Better Place to Live.*
McCamant & Durrett, *CoHousing.*
Urban Land Institute, *Mixed Use Development.*
Ayres, *Crossing the Energy Divide.*
Murphy, *'Beyond Fossil Fuels.*

# 14

# TARGETS

---

**Raw Material Sourcing | Fabrication Targets | Consumption and Discard Targets | Free-Flow Energy Capture | Doubled Fossil Energy Use Efficiency | Progressive Fossil Carbon Taxing | Reduced Nitrate Manufacture | Transport Demand Management | Soft Pathway Fertility Decline | Compliance and Penalisation**

---

The previous chapter 'Goals' set out policies for sustaining a wholesome habitat long term. This chapter 'Targets' pragmatic conduct; those affirmative, direct, material and notably personal and community adaptations taken to support a balanced lifestyle which secures a resilient biosphere. It runs on from the National Goals, Community Objectives and Household Responsibilities which concluded the previous Chapter 13.

---

Appendix C — Exemplars from Oceania VI
Melanesia-Polynesia: Pragmatic Adequacy

---

Environmentally conscious lifestyle manuals of several kinds exist (*see* References). Mostly they are Trans-Atlantic or Australasian. None of them has moved humankind with certainty to correct the quantum problem, the tsunami of wealth-driven, oil-fuelled consumerism for an increasing human mass.

Throughout, these pages have identified fossil fuel and nitrate waste excesses as the principle discard factors up for prescribed adjustment. The recommendations are mindful of the observations quoted earlier from Jung, Marx and Curzon identifying our aptitude for evading reality, avoiding problems, and procrastination. They are observations to be also weighed against Popper's dictum about the near impossibility of choosing between ends: choosing between resource consumption (hedonism?) and ecological balance (virtue?) is proving problematic. Catch-21 indeed for humankind in this 21st century![173]

The professions with which I am most familiar — development planning, architecture, engineering, landscape design, regional development, urban planning — work from 'the whole to the part'; and that approach has been adopted here.

---

The 'big picture' concern of the earlier chapters has been consumer ethics; now the focus moves to a consideration of consumer conduct. Where previously the gaze was outward, it is now inward toward the point of consumption; the community, family and personal setting of the homeplace the workplace the marketplace and the schoolplace. It concentrates on the foreground to our lives.

---

Understanding the *'Why?'* and the *'What to do?'* need for improving change, the repeated mantra, is important. The key to this change is big-horizon adjustment to the economic, demographic, carbon-consuming and nitrogen-fixing fundamentals. In terms of targeting it comes down to which style of technology, what scale, which to include, what to avoid, and which modes of transportation and lifestyle? From bottom-to-top (being more effective than top-down) it's a matter of getting to be 'really clever and really right'.

---

At national and community levels the toolbox contains many good-practice instruments. These include: enforcing the polluter-pays-principle, poleaxing fiscal buccaneering, fashioning preventions, adaptations and incentives, and installing Input-Output tracking.

---

Good-practice also includes eco-information and eco-education, demographic constraint reforms, cap-and-trade to some extent, emission controls taxes and penalties, and adhering to national and local initiatives; all the while seeking-out and heading-off every environmental malpractice.

The challenge is for these practical specifics to be taken up, differentially, by rich-world and poor-world peoples, and to maintain the output of GlobalGood information and findings (Chapter 10) to that end.

Integrating conservancy criteria into local and personal life — Karl-Henry Robert's *Natural Step* — comes down to how we live our daily lives, how we as individuals and householders work, play and behave, and how we consume and discard. This applies also to the poorer four-fifths of humanity, although it is clear that this sector is already unselfconsciously on this pathway through inherited circumstance. In order to survive in the longer term we have to learn to live contentedly with our local situation; to move around locally under our own steam, to produce food and energy locally, to work and school locally, to seek recreation locally; and to waste-not and trash nothing reusable.

Of seemingly intractable and continuing concern is the environmental damage perpetrated by the over-wealthy in excess of the nations-of-wealth average, which of course vastly exceeds the global average. Surcharging (progressively taxing) carbon purchases at the point of pick-up is the key to inducing lessened carbon use.

A problem here is that fossil carbons would still be over-consumed by the wealthy, indulgently paying the financial penalties for profligate carbon uptake. A drastic poleaxing of binge carbon consumption through a combination of allocation procedures, taxes and penalties, is socially and environmentally justified.

'Targets' in this chapter have been worked-out differently from other alternatives given in the References. The main distinction is that the format adopted here embraces those of wealth as well as poverty, as well as those from the political 'left' and the political 'right'. This recognises that we are all obliged to return to living again in a 'softer' way, according to the modes and styles dictated by the limits of waste absorbtive resilience.

What follows connects pragmatically with the three previous chapters (Economics and Demographics, Resource Conservancy with Waste Conservancy, and Goals) set down in this chapter as 'Targets' arranged and examined top-down:

- **Raw Material Targets**
- **Fabrication Targets**
- **Consumer Targets**

Permeating all three groups is that much repeated exhortation to 'manage-up efficiency' and 'wrestle-down inefficiency'.

The first of the three groupings addresses the take-up of raw materials. This is mainly about controlling resource over-supply (notably mineral carbons), which is the supply sysyem to curtail if a turnaround lessening of carbon gas effusion is to be achieved. Currently there are paltry reduction-specific regulatory guidelines. In this situation supply simply meets energy demand at the highest price the carbon market can command while it, the market, seeks to further increase consumption. This is exacerbated by an institutional inability to ensure that carbon consumers meet the externality cost of their free-to-air carbon gas discharges.[174]

## Raw Material Targeting

Raw Material Targeting involves the controlled management of feedstock supply. If, in an environmentally balanced future, 'less' is to result in 'more' then it is specifically a curbing of the uptake of fossil fuels, atmospheric nitrogen, natural forests and ground water by a lesser population, that will beneficially stabilise the climate. This requires a massive 'sharp end' curtailment of resource sourcing, particularly on the part of the urban living consumers who constitute the waste-producing majority. A feedstock complexity is that sourcing operations are dominantly away-from-sight open area activities carried on 'somewhere else' and 'somewhat beyond' understanding and reason.

Plantation forestry undertaken in open territory offsets a little of the excess carbon dioxide emanating from manufacturing processes, and attracts approval; but any absorptive benefit stops when planted tracts are harvested. One permanent forest protection alternative involves the contra-flow of richer nation finance to poorer heavily forested nations with a view to purchasing and protecting tropical rainforests, the really effective carbon gas absorbers. This is put out through the United Nations as the Reduced Emissions from Deforestation and Degradation (REDD) initiative. A major flaw here being the use to which pay-to-preserve moneys are put; another is the reversion to forest clearance when the cash-flow stops (leading to the likes of soya planting in Brazil, oil palm tracts in Indonesia, groundnut cropping in Tanzania); another is compliance and enforcement in nations bedevilled by corruption.[175]

Carbon Companies have been established to purchase forest tracts — the Carbon Rush — securing carbon credits through the preservation of indigenously forested landscapes, mainly in poorer nations. Their underlying motive is almost the reverse of conservancy: it is to secure pollution offsets for dirty industries in wealthier nations, for profit.

The overall point to make here is the need to ensure that indigenous forest loss, in all tropical and temperate rain forests and throughout all Mediterranean-style terrain, is countered permanently. Overall forest cover and planetary carbon dioxide absorption capacity must be increased through the enforcement of between-nation protocols; not through hope based on the futile falsity of carbon company offset-purchases.

The $CO_2$ absorptive capacity of conventional horticulture is relatively insignificant; and is cancelled out by the methane produced by ruminant livestock. Every retraction of pastoralism (principally through lesser meat food selection) achieves overall improvement to atmospheric balance.[176]

Also important is retention of topsoil and aquifers. This is achieved by denying soil degradation and regulating water extraction, adapting to the rate of natural water replenishment and soil nutrient retention, and the use of some future-proven not-for-profit GM plant materials. Permaculture, involving local food fibre and timber harvesting is socially responsible and environmentally beneficial.[177]

Mechanised feral fishing (particularly bottom trawling, long-lining and seine netting) is an environmentally irresponsible way of providing food because it is undertaken by individuals driven to outdo their competitors, setting-out to gain from beating others in the race to exploit a 'free' common resource. Rapacious selective-species fishing is pretty much the norm; with chasing about the Antarctic Ocean in whale catchers and the targeting of toothfish about as obscene as protein food supply gets. On the other hand fishpond farming (as with tropical tilapia and temperate carp), and low-impact inshore fish farming and fishing from small craft in conformance with managed quota, is environmentally viable.[178]

---

**Communities must avoid lumbering along with ever-worsening carbon gas effusions; to institute direct ways (alternatives, sequestration, quota, taxes, penalties, incentives) to reduce the rate of carbon consumption; specifically, for persons-of-wealth, from annually around ten-plus tonnes to two-minus tonnes per person.**

---

This part of the discourse tracks back, inevitably, to the mining of fossil fuels; the oft repeated problem being a supply of oil, gas, coal, tar sands and oil shale in excess of the planet's capacity to handle the offload emissions. Over-supply of mineral fuels is linked to an over-sized and over-consuming population not made to pay for its free-to-air discard.[179] It is imperative to emphasise energy capture from freeflow sources; solar energy capture, river hydrology, tidal kinetics, geothermal and wind power generation. Climate stabilisation requires reducing and substituting for the uptake of mineral carbon feed stock, along with a doubling in the efficiency of future fossil carbon energy use.

With natural energy capture devices there is also the need to factor in the carbon sequestration costs of manufacture, installation and maintenance. For example, a domestic photovoltaic-and-battery system costs more to manufacture and install than a solar heated hot water system, with the latter the more efficient, longer lasting, and effective domestic energy producer.

Consider another example: mega-hydro dam installations are hugely demanding of energy during construction, relative to small-dam and run-of-river installations, at a rate of return to weigh-up in terms of the respective energy yield. This rate-of-return factor looms large in the context of biofuel and biodiesel manufacture. Conversion of freeflow solar energy through photosynthesis into a carbon feedstock, is surely a plus; but at what rate of return after factoring-in the labour and mineral-fuel inputs involved in the supply of the biofuel?

---

Beyond, and an improvement on plant biofuels, is the prospect of anaerobic (microbial) digestion of biowaste. Putrescible tallow, guts, raw sewage, farmyard manure, organic garbage, woody biomass and smokestack carbon dioxide in; liquid and gaseous fuel, plant fertiliser, and water out.

---

The symmetrical beauty of microbial digestion is that there is only a small amount of toxic residue to cope with as true waste. From such muck, there surely can accrue much environmental redemption. The technology is undergoing development, the prospects are promising.[180]

Mineral carbons from finite lodes remain the dense 'fuelling drug' of convenience and choice simply because mineral carbons still cost less money per energy unit delivered, with the usage largely untaxed, and the toxins off-loaded as forsaken and forgotten freebees (an externality cost to the environment).

Worse, with the uptake of heavy oils, tar sands and oil shale there exists sufficient future carbon gas effusion to compromise worthwhile human existence as we know it. The inevitable agony may be masked with palliatives; but in the longer term human welfare can only be sustained through a lesser use of mineral carbon.[181]

Here's a seven-way matrix of concomitant carbon reducing actions:

1. An across-the-board halving of current mineral carbon use, driven by a tax-and-regulate system.

2. Incentive and reward systems for doubling carbon-use efficiency.

3. Pricing and penalising the downside externalities associated with excess carbon gas emission: If you use legitimately, you pay a simple progressive tax; when you pollute aggressively you clean-up, and face a penalty.

4. Enhance $CO_2$ capture through plantation forestry and permanent re-aforestation, along with techno-sequestration.

5. A fifth leavening option involves the previously mentioned, but yet to be perfected, large-scale digestion of all manner of biodegradable waste for the production of biofertilisers, combustible gas and water.

6. The potential of extensive and ultimately more cost-effective installations of solar, wind, moving-water, heat-exchange and geothermal energy capturing devices.

7. A displacement, short-term option, is uncarbonised energy from new era fail-safe nuclear fission reactors.[182]

The lower intensity criteria — **Clever and Wise**, **Sustainable and Soft**, **Positively Proactive** (from the previous chapter) also apply.

When less fossil fuel is mined in situations of increasing demand, the price of oil (and all else) goes up. This demand-supply and wants-needs situation for a base commodity is so intertwined, mesmerising and distracting it screws-up the ecological perspective. An overarching point to make is that because fossil fuel reduction is the key essential for stabilising human survival, the fixing of carbon pricing, along with taxing and penalisation for wrong use, is fully warranted. Restrictive carbon allocations, progressively taxed and prescriptively enforced, is the most direct and effective way to limit damage to the biosphere.[183]

From early geography lessons we know how important it is to shorten the triangular distance between resource sourcing, processing, and consumption. The optimum is local mining and quarrying, local energy generation, local timber milling and local food production; in effect energy, mineral, food, fibre and timber supply patriotism. Overarching is that vital charge; to ensure that the emission and discard rot gets arrested through taxed and controlled mineral carbon and nitrate supply.

The recycling of carbon-infused end products is of course worthy, but too after-the-event to be of practical utility — and of course there is no redress for nitrate discards accumulating in the hydrosphere.

**Fabrication Targeting**
Fabrication Targeting is second-ranked after raw material sourcing. Manufacturing, construction and food processing activities are subservient to, and their material supplies are monopolised by, raw material supply.

A feature of building construction, hardware manufacture, agriculture, and food and fibre processing is that all these activities are dependent on mineral carbon inputs. The chemistry of cement manufacture, for example, generates a weight-for-weight output of carbon dioxide as a major proportion of the end product; and is so emission's effusive that it daunts most cap-and-trade advocates.[184]

Industry absorbs around a third of the mineral fuels and lubricants produced. Variable-speed machine drives and heat exchange capture, along with other energy use efficiencies, natural lighting and hoped-for $CO_2$ sequestration, contribute to a reduction of carbon uptake and emissions. Pricing and taxing systems for industry, as with cap-and-trade, are also helpful within any consistent economic climate; but again neither sustainably, nor lastingly, when economies crumble and politicians buckle.[185]

The worker input to extraction, manufacture and construction in wealth economies can be further rationalised through an emphasis on fewer, possibly longer, working days within a shorter working week. The lesser amount of formal pay cheque work available in a less carbon-fired future becomes socially challenging; which opens-up and relates to the possibility of more local and labour intensive construction, food supply, schooling, energy capture and waste disposal arrangements.[186]

Overriding, within Fabrication Targeting, is the 'efficiency' mantra; to minimise mineral carbon uptake and nitrate use, avoid (or capture) toxin output, use local labour, and introduce softer cleaner and waste-avoiding practices. Efficiency, (contraction-and-convergence) is a key emphasis, education and information a key conduit, with input-output analysis providing feedback and guidance.

The principles for leaner fabrication procedures also apply to farming and timber producing activities. They are: to use smaller and more efficient machinery less, to achieve rain fed rather than irrigated product, to avoid the use of nitrate and phosphate derived fertilisers and oil-based insecticides herbicides and fungicides; and to sell and trade goods locally.

**Consumer Targeting**
Consumer Targeting is a community, family and individual derivative; a follow on from the previous Sourcing and Fabrication targets.

Most of the consumer maxims which follow are for individual, household and community attention — the workstation, the schoolplace, the recreation and entertainment venue, and the shopping place — remembering that the overarching goal is the attainment of sustained environmental certainty on humanly beneficial terms. Consumer Targeting is largely about eco-manners; taking its cue from an ethos that lines-up with the need to remain within the sustainable limits dictated by the resilience limits of waste absorption.

With transport and communications the emphasis is less of the former and more of the latter; collectively strategised as TDM (Transport Demand Management). Efficiencies in transportation can be achieved in myriad ways, principally by cutting out non-essential transportation.

This involves avoiding unnecessary journeys (near-empty trucks passing each other with light loads, and the solo-occupied commuter car); the localised sourcing of raw materials and food supplies; and giving-up the fly-to-and-back holidaying. And, as noted earlier, a significant reduction in transportation movements can be achieved as a result of a MUD (Mixed Use Development) urban strategy along with improved communication — wirelessly and electronically — contributing to a reduction of unnecessary personal movement.

The Victoria Transport Institute (Australia) targets four categories of Transport Demand Management:

- Policy reforms
- Improved Transportation Options
- Incentives to Avoid and Reduce Driving
- Improved Landuse Planning

For further details from this significant analysis consult the endnote.[187]

Energy savings and benign modes of energy generation can be introduced around the workstation and homeplace by installing natural heating and cooling, insulation, solar energy capture, and through the adoption of 'smart metre' energy management. Consumer adaptations and prevention includes all the waste avoidance edicts — reduce, recycle, reuse, rehabilitate — along, of course, with resistance to unnecessary purchasing and packaging. These are all strictures of less relevance to poorer people who don't generate much in the way of hard waste for disposal. Poor people are also adept at composting organic waste.

Environmentally sound design, and the use of locally sourced construction materials assembled by local craft people, contributes to reduced mineral fuel consumption.[188]

Adopting the GreenTick logo assists thoughtful purchasing by environmentally conscious shoppers; notably through data showing the energy embedded into, or carbon dioxide produced by, a product. More advanced, and gaining in scientific credibility is the designation on products of an 'equivalent carbon dioxide emission' ($CO_2$eq) reading. Couple this provocative information to kitchen gardening, waste recycling, personal mobility and solar energy capture, and the individual contribution to environmental gain becomes tangible.

The urban basis to the litany of consumer efficiency has its rural reflection: in waste avoidance, improved efficiency of lesser used mechanisation, reduced chemical controls and fertilisers, wind, hydro and solar energy capture, diversified and mixed farming, and farm gate marketing. In open-area landscapes it is vital to control the extraction of groundwater and retain surface runoff, to avoid the exhaustion and erosion of soils, and to maintain tree cover.

---

The overall emphasis is on producing locally from nature, neither taking out more than a landscape can support, nor generating damaging toxins and runoff wastes in excess of those the biosphere can absorb.

---

The practical objective is to lower mineral carbon and fixed nitrate embodiment into food and fibre supply, all the while conserving soils and aquifers for stable and sustained food and fibre production.

This review of Targets needs also to address the delicate and divisive matter of population restraint through planned fertility decline; the matter of eventually getting birth rates in alignment with death rates. The issue has been already backgrounded in the Part A: 'Limits of Resilience' chapter on 'People and the Environment', and in the Part B chapter on 'Economics and Demographics'. Targeting birth-limiting practices divides into a consideration of repugnantly *hard* (or harsh) and socially *soft* (or benign) social variables.

The *unacceptable options* include spectacular, largely involuntary, and distasteful adjusters — epidemics, pandemics, plagues, disease, pestilence, cyber-attacks and natural disasters. To these are aligned the hateful and repugnant visitations of war, genocide, ethnic cleansing, enforced displacement, and abortion on command. Socially acceptable and effective at the consumer level are the *soft pathway options*. These are linked into educational programmes; settled on through individual awareness and timing, induced by agreement, and supported with community leadership and understanding. Included here is societal and community tolerance for non-productive marriages, gay and lesbian childless partnerships, childless singleness, and orthodox one-child or two-child parenting.[189]

Add to these, social reinforcement and official endorsement. Surrounding the soft options, available throughout every community, should be access to cheap (or free) advice and help with contraception, sterilisation and self-choice abortion. Population reduction can, to a degree, be 'willed' according to strategy; but only gradually, in personal soft-option terms.

The core real world objective, overall, includes a halved reduction of mineral carbon mining, a doubling of energy use efficiency, and an increase in forest cover — all in association with fiscal and demographic stabilisation. Pointedly, mineral carbon extraction and nitrate manufacture must be quota-mandated and progressively taxed in accordance with mainly non-market decisions and protocols. The biospheric limits to growth and absorptive capacity are there to be respected, and prescriptive means and methods are there to be fashioned.[190]

Compliance has to be certain: a cultural imperative. One strike and you suffer personal and corporate consequences, with you and your agency's producer-consumer quota curtailed or denied; and in the worst-case situation of ecocide the closure of your enterprise and exemplary penalisation. What is needed is a jurisdictional commitment to environmental justice, the moral-with-pragmatic basis for ensuring a socially worthwhile and environmentally stable future.


## FURTHER READING

Lovins, *Soft Energy Paths.*
Steffen, *World Changing.*
Schaeffer, *Soft Living Sourcebook.*
Charter & Tischner *Sustainable Solutions.*
Goodall, *How To Live a Low Carbon Life.*
Holgrem, *Permaculture.*
Mollison, *Permaculture.*
Nye, *Soft Power.*

# 15

# PART B: OVERVIEW
# & SUMMARY

A humbling perspective has come my way from Roderick Smith of Imperial College London. Each doubling of global economic growth during the 20th century consumed finite resources equal to those consumed over all previously recorded growth-doubling era. Clearly, our finite orb cannot support another redoubling of consumption for a redoubled consumer mass with unrestricted access to the vast remaining fossil carbon reserves.[191]

One hundred and fifty years ago John Stuart Mill, in his day every learned person's radical free thinker, foresaw the problem of leapfrogging expansion, and willed society to attain a 'stationary state' which exhibited accomplishment, but without growth; what in contemporary terms we would now style as living sustainably. It was his hope that from an equilibrium established along these lines, the problems being created by 19th century industrial growth would be self-limiting and correcting.[192]

The gulf in time, and difference in approach, between JSM's desire for a stationary state, and Rod Smith's recent account of the haunting consequences of made-up-money and hyper-consumerism, are considerable. What each established, from widely separate standpoints, is that humanity cannot consume and expand indefinitely. Mill foresaw and Smith foresees, that growth based on increased carbon supply, and free-rider discard beyond the global capacity for waste absorption, will not endure.[193]

Of course the World will not come to a standstill but, simply put, we cannot have climatic certainty, *and* perpetual economic growth, *and* a constant increase in population.[194]

The first climate change Conference on the Human Environment at Stockholm (1972) was followed by a succession of other gatherings leading to the energetic Agenda 21 agreement at Rio in 1992. Then, after a slew of side-shows at several other exotic locations came the Rio+20 (2012) Earth Summit, a non-event sporting a vague aspirational logo 'The Future We Want'. Over much the same period the rich-list G7 nations (morphing into the Group of Eight really rich nations: plus the G20 attachment) produced a parallel bevy of reports.[195]

Streams of reportage infer the now blindingly obvious for a finite planet: that the obstacle to environmental stability is too many of us, too much conjured wealth, too much consumption hence too much waste, self-interest too dominant, governance possum-paralysed.

## UN CONFERENCE OF PARTIES

What is evident from the multitude of climate change (COP) events is that the leader of each delegation carried around in their head two sets of numbers, one bothersome, the other worrisome. The bothersome number being their nation's carbon-guilt ranking, how relatively badly they were placed in the binge-emissions league. The worrisome number being their nation's ranking, and its relative obligations, in the binge-fiscal league.

Calculators at the ready, delegation leaders get set to 'negotiate'. These negotiations are only marginally about the pivotal pragmatics of carbon-use reduction and protecting rainforest carbon sinks. What they are really about is how much conscience-money delegates from the gross polluting top-half of the fiscal league will have to stump up to address climate issues – on which they were mostly ambivalent!

From the perspective of poorer nations the negotiations concerned the proportion of that money ticketed to filter through to them to assist their adaptation to climate changes – mostly not of their making!

The 40 years between the Stockholm 1972 and Rio+ 2012 events was a time of unprecedented increase in rates of unbacked money supply and an expanding population buying, consuming and waste producing. The fiscal heist, in particular, incited the plunder of resources that generated the waste discards that then induced climate change and sea level rise

After the mid-2008 financial collapse the resulting recession profiled as a positive blip for the environment because of a briefly dented consumption of oil and a lessened nitrogen fixing. Then followed a reflex monetary grasp, leading governments to bail out the fiscal mistakes. This encouraged an officially sanctioned, hugely massive, and furtive con 'quantitative easing' through an electronic 'conjuring' of money.

Let's be clear here: tipping shed-loads of hey-presto cash into the consumer maw is no way to avoid overloading the biosphere with toxins; or, indeed, to avoid another round of hyperinflation and economic collapse. The time has come to rein in, stabilise and regulate monetary services; time for a mode of economic management that cooperatively reconciles the biospheric and fiscal value systems. On this I wish and I hope; yet I also admit to a despair; that inertia will overwhelm us. In short: we can either pursue consumer and population growth for a few decades, or stabilise our habitat for the long haul; but, most certainly, we cannot do or have both.

Emissions curtailment is led by environmental scientists pleading for direct carbon gas and nitrate reductions. Their efforts are paralleled by macro economists striving to bulldoze a way through with a surge of 'printed' money to correct and buy-out mistakes. And then there are the demographic specialists seeking smaller, better balanced, and happier populations.

None of these approaches (with 1,500 Environmental Science Studies courses offered in the USA alone) can work in isolation from the others, or separate from the political power base. Point the scientists, economists and demographic advisors in the same direction, link their understanding of boundaries and limits to a planning advisory service with political power, and the function of environmental justice and biospheric stability coupled to prevention and adaptation has a good chance.

---

Avert embezzling the future quality of the living biosphere. This is principled and wholesome. It calls for a prescriptive approach incorporating an all-people's, all-nation's, pan-political socio-legal objective, a new style of plan-led, environmentally-governed, capacity building, and a stable monetary system. It is at once morally worthy and generationally just.

---

## THE CORE PREVENTION AND 'ADAPTATION' MARKERS

1   Acknowledge that an identification of the planetary boundaries of resilience to waste absorption is a deep-time human obligation.

2   Entropic and climatic stasis requires humanity to: Reduce carbon uptake, nitrate use, and natural forest destruction.

3   Establish a steady state economic (SSE) system. Manage socially acceptable fertility decline.

4   Inform sceptics, set right the deniers, correct the misinformed: Suffuse education curricula with sustainable and conservancy values.

5   Commission an open-input independent GlobalGood super-computer facility; collate, interpret, trend, forecast, update and disseminate all manner of biospheric information and analysis.

6   Curtail fiscal buccaneers; outlaw fiscal alchemy. Let us not be fooled again by fiscal fraud masquerading as competence.

7   Reset the moral compass to a 'new ethics' and 'new politics'. Tertiary graduates to swear an oath of lasting social service.

8   Reproduction: 'stop at two' and support those who 'stop at one'.

9   Resource Conservancy with Waste Conservancy. Promote Efficiency and scotch Inefficiency.

10   Introduce 'Input-Output' procedures; along with 'Contract and Converge' and 'Polluter Pay' protocols.

11   Actuate Pragmatic Environmental Justice through: Small Scale Guidelines, Ecologically Sustainable Guidelines, Technologically Appropriate Guidelines.

12   Avoid, reject and punish environmental assault and criminality.

# Live Lightly Locally

"Two roads diverged … and I,
I took the one less travelled by,
and that has made all the difference."

— ROBERT FROST

# APPENDIX A: CONSUMER BEHAVIOURS CONCEPTUALISED

Anatomical advantages, a larger and better-wired brain and the 'selfish gene', has led humankind to biospheric disdain and within-specie aggression. Jacqueline Feather, Clinical Psychologist modelled this for me as a four-part social pathology.

| EMOTIONS | THOUGHTS |
|---|---|
| 'Negative Emotions' such as anxiety, greed, emptiness, envy and fear; also 'Feel Good' emotions such as the exhilaration associated with status, wealth, display and conquest. | 'Persistent Thoughts' relating to keeping up appearances, to impress; looking out for self and family, maintaining control and power. |
| **SENSATIONS** | **BEHAVIOURS** |
| Physiognomic 'Stress Sensations' such as elevated heartbeat, sweating, muscular tension, panic; also 'Satisfaction Sensations' (triumphalism) following consumption, sex and euphoria, domination, and gratification. | 'Questing Behaviours' which drive individuals to purchase, consume, use and discard, squander, accumulate, hoard, demand, justify and dominate. |

Her construct illustrates the way possessive individuals strive to dominate and feel better through a self-interested and environmentally disastrous 'freedom' to plunder, gorge and hoard. This has become accepted as the so-called normalcy-bias, allowing wealth accretion to emerge as the pinnacle expression of our character.[196]

The construct also profiles how 'hedonic wants' are in excess of 'sustainable needs'. The easiest individual failing to rein-in involves our **BEHAVIOURS**; toughest to influence are our **EMOTIONS**. Other aspects of our mental and physiological makeup, our **THOUGHTS** and **SENSATIONS**, make it difficult for us to recognise and disconnect from habits, which implicate us all in climate change (accumulating wealth, the hedonic treadmill, pathological consumption).

Human beings are not intrinsically virtuous; we are indeed 'too smart for our own good'. A characterisation of behaviours, as wavering between 'wicked' and 'benign', connects to the above.

| WICKED | BENIGN |
|---|---|
| Mistrust | Love |
| Deceit | Truth |
| Ruthlessness | Tolerance |
| Disorder | Order |
| Footloose | Belonging |
| Chaos | Balance |

The **'WICKED'** column is sourced from Rittel and Webber *'Dilemmas in a General Theory of Planning'*, 1973. The **'BENIGN'** column is based on Maslow's 'Hierarchy of Needs' (*Toward a Psychology of Being, 1968*).

Also, here's a gem of prescience from Tim Jackson at the UK Sustainable Development Commission (2010):

"We spend money we don't have, on things we don't need, to make impressions that don't last, on people we don't care about."

# APPENDIX B:
# WARNINGS FROM OCEANIA

## OCEANIC WARNING I
## KIRIBATI: CLIMATE CHANGE & SEA LEVEL RISE

Before European 'discovery' the population of the remote atoll nation of Kiribati, comprising more than thirty islands scattered over six million square miles of ocean, had access to the merest of resources and food supply. Brackish water was sourced from a rainwater 'lens' six feet below the dry sand surface, fish and shellfish provided protein, the coconut and a rough starchy taro provided a vegetable balance, pandanus fibre was used for clothing and thatching.

Shortly before independence I was commissioned to prepare a development plan for the capital centre on Tarawa. Being there, doing the job, was a sublime experience; pure lagoons, picturesque palm-covered islets, joyous and carefree people. Today that has all turned to custard! Tarawa is now overpopulated, palm-denuded and rubbish-infested; with the critical life-sustaining brackish water 'lens' salting-up as higher and yet higher spring tides cover more and more of the precious sandy domain.

Sea levels are said to be rising 5.6mm a year ('*certainly*' 3.7mm annually '*and increasing*' from 1993 to 2003 according to a 2007 IPCC Report). That rise is linked to global warming, which is linked back to carbon gas emissions, which is linked back to ever-increasing mineral carbon burning. The net result is obliteration of a nation right before the world's television cameras. Kiribati (and Tuvalu, Tokelau, Maldives) profile a clear global forewarning: greenhouse gases smother, hotter temperatures melt ice, oceans rise, and populations get displaced.

India, Japan, Indonesia, Vietnam, Bangladesh and the USA each have more than twenty million population living within a band ten metres above sea level; China has almost eighty million people living within that band! Particularly vulnerable are low lying atolls, much of Bangladesh, a large part of Vietnam and The Netherlands.

## OCEANIC WARNING II
## SOLOMONS: THE 'CARGO CULT' DELUSION

An inference and a message for all consumer peoples can be drawn from the post WWII 'cargo cult' consumer delusion of Solomon Islanders following the 1940s war in the Pacific.

The American counterattack against the Japanese on Guadalcanal came over the horizon, unannounced, with Liberty Ships run straight onto the beach at the edge of Henderson Airfield. The place, Honiara, came alive with ships and planes disgorging personnel, jeeps, trucks, materials, food and toiletries in abundance. Then the Pacific War moved on and the servicemen with it, abandoning all the buildings and stores along with an infrastructure of roads and vehicles.

The indigenous population never anticipated the booty delivered fortuitously to Guadalcanal; and it was something of an aftershock to them when supplies were no longer forthcoming. So, desiring the return of their American benefactors, the local people set out decoy planes, fashioned from rattan, to attract the silver planes back; instituting a cargo-cult which bemused western observers.

Following on from the indigenous people's reaction was the response of the colonial-ruling British on their return after the war to readminister the Solomon's. On Guadalcanal was everything required for a new-start capital; a centre of government fashioned out of the infrastructure and buildings left behind by the Americans. I was asked to prepare the plan for a new capital centre, Honiara, incorporating the abandoned infrastructure and materials to hand.

And that earlier inference? Keep in mind the Solomon Islander disappointment; the failure of cornucopia to keep on coming over the horizon from distant lands, continuously and endlessly. Reflect on this

Solomon Island cargo-cult parable as a portent for the future non-delivery of consumer goods to all corners of the earth.

# OCEANIC WARNING III
## SAMOA AND SAMOA: RICH & POOR

In the early 1980s I assisted with the preparation of a Country Report for the Pacific Island territorial congress known as 'American' Samoa, the smaller sized and less populated cultural extension to fully independent 'Western' Samoa. The project was notable for the openness, frankness, enthusiasm and youthfulness of the personnel sent from Washington to administer benignly to this federal outrider.

Stateside administrators, dispatched on yearly contracts, judged that in terms of the criteria established for regions of mainland poverty, America Samoa warranted Federal beneficence. Based on those criteria most of the population was given income support, housing grants were provided, infrastructure was installed; and because the turnover personnel from the States could not get up-to-speed with the Samoan language, every household was allotted a television set so they, the locals, could get up to speed with the American-English language!

Gratuitously funded and culturally indoctrinated, the indigenous people spent their assistance cheques on fast food and sweet cola, becoming morbidly reliant and bodily obese. Served a distracting diet of Stateside television, and Stateside culture, traditional Fa'a Samoa withered. Financially American Samoans became relatively rich; parallel in time they became culturally impoverished.

Their financially poorer, ethnic and linguistic cousins from independent Western Samoa are, by contrast, healthy, well fed, and culturally secure. The 'warning'? Being patronised, even by a well-to-do and well-intentioned outsider, is culturally dispiriting and physiologically degrading.

# OCEANIC WARNING IV
# BOUGAINVILLE: CORPORATE GREED

In the mid-1970s the Panguna Copper Mine was in production on Bougainville, accompanied by rumblings of discontent from the indigenous Melanesian population. I made a sneak visit, hitching a dump-truck ride up to the mine, through the guarded gates into the administration complex, where I found my way to the PortaCabin office of the Environmental Manager. Surprised, he accepted my tourist-passing-through explanation, and showed me around.

Everything, from the vast hole where the volcanic plug had once been, to the mining machinery, was gargantuan; except the already half-full slurry dam below the processing plant. A year later the toxic slurry began to overflow, killing stream life in the Kawerong River all the way to the sea; then in both directions along the adjacent coastline.

Pitiful compensatory cash payouts were made; and I was told later that there was a time when claimants had an arm stamped with an indelible marker in exchange for a $100 bill! Disquiet festered, revolt followed, civil unrest erupted, mining stopped: life ground to an uneasy halt. A blanket of destruction, decay and death engulfed the island.

There had been an illegal confiscation of land, the Royalties Contract was fraudulent, there was unsympathetic government connivance from the capital Port Moresby (mine revenues underpinned the Papua New Guinea economy); and to top-off this infamy there was that wholesale poisoning of river and coastal waters.

All along, the propaganda machine side-stepped the truth; setting off the specific environmental destruction which brought about a descent-into-hell for the whole island; an industrial and human disaster (backgrounding the stellar novel by Lloyd Jones *Mr Pip*, 2007) totally destroying human life as previously enjoyed on Bougainville; a parable of corporate greed for the rest of humankind to heed.

## OCEANIC WARNING V
## FIJI: DISCORD & CHAOS

Easter 2009. Revisiting Fiji with a study group I reflected on Fijian politics before Independence (1970) and after; following the first four — or is that five? — *coup d'état*: the 2008 event, with a 2009 add-on going a step further, discharging the judiciary!

Fijian overthrows are sullen affairs, with only a few people identifiably maimed, giving rise to accounts of closet detention and roughing-up in the military barracks. What has changed noticeably over time is the people's attitude; the 'paid to smile' subservience to visitors and the desperate 'buy something from me' poverty.

There is also the decay of meso-order at local government level; with centralised military power inducing local slippage. Of interest was the outcome of my previous (mid-1960s) planning work. As a consequence of centralised military control it was apparent that local government was pretty much in abeyance; now nothing got attended to except in response to directives coming straight from the military, devoid of democratic content.

Of course, all democratically elected government is flawed — proof being an inability to constrain the fiscal beast in nations of wealth, and a contrasting failure to help the oppressed in poor nations. Yet democratic government remains the somewhat baroque least-bad conduit available for environmental prescription.

Dismembered and chaotic military control (Fiji and previously the Solomons), and monarchy (Tonga), are omens from Oceania for the rest of the world to heed. Undemocratic misrule engenders corruption and disorder.

## OCEANIC WARNING VI
## TARAWA: POPULATION EXCESS

Tarawa is the islet capital of the remote Kiribati Nation (Oceanic Warning I), seventeen islet atolls including the large Christmas Island. Tarawa is also known to North Americans as the scene of a major WWII battle against the Japanese, with horrendous American loses and no Japanese survivors.

In 1968 I prepared a development proposal for the Kiribati capital; a planning no-brainer as the settled area was a two-dimensional beading of connected islets averaging two hundred metres wide strung out over thirty kilometres, with a port at one end and an airfield at the other. The only potable water source available involved careful abstraction from the 'water lens' exposed in the bottom of brackish trenches known as galleys. These were dug along the centreline of selected islets set aside as water reserves.

The per-person rate at which this brackish yet drinkable water could be extracted was set at a maximum of ten gallons per lineal foot of galley per day; for it had been established that if this rate was exceeded, seawater would filter into the water lens, and human life on Tarawa would become untenable. On this basis the maximum allowable water abstraction (totaling 70,000 gallons per day in the dry period) divided by the minimum daily consumption per person (calculated as a minimum six gallons per day) worked through as a safe allowable population figure of 11,500 for Tarawa.

Attracted to Tarawa as the Nation's capital, with possibilities for employment, education or onward migration, the population in 2011 was around 55,000! Keeping in mind its remoteness, and that sea level rise is starting to swamp the land, it can be understood that population pressure on Tarawa is extreme. For the rest of the world to heed: Tarawa illustrates the hardship of unrelenting and unrelieved population overload.

# OCEANIC WARNING VII
# MALIMA: TECHNO-FRAILTY

Fiji 1965: an opportunity came to spend a few days marooned, voluntarily, on a remote uninhabited atoll — Malima in the Explorer Group.

While the SS *Andi Keva* idled outside the reef my wife and I, with our gear, were taken ashore in the cutter on the understanding that we'd be picked up three days later, after two nights 'cast away'. With the *Andi Keva* hooting good-bye, we were all alone, isolated in a blue-green and golden world, almost out of sight of other land.

Malima's most interesting feature was its hillock; a rare vindication of Darwin's crafty explanation for ringed atolls; where reef-fringed land had slowly disappeared below the waves, leaving behind a coral circle. We set up camp at the base of the hillock.

Around ten o'clock in the evening the clear sky to the north was suffused with a vast pink and orange glow. There was no sound. We knew there was no island in that direction for several thousand miles, so were stumped for an explanation. Fearing a volcanic eruption and tsunami we climbed to the safety of the hillock. An hour later the glow faded. The next day was spent exploring our small domain, wondering often about this phenomenon.

Late on the second evening Malima was hit by a storm; strong wind followed by heavy rain. The pup-tent was not much use. Cold and miserable we waited for the dawn, and the return of the *Andi Keva*. All day, no sign? The wind dropped, but the rain kept up. The following morning, after an extra unscheduled night, wet cold and now a little scared, the *Andi Keva* came up slowly and belatedly over the horizon. Immediately we sought news of the vast northern glow on the first of our three anxious nights? What had happened? The explanation: a nuclear device had been exploded in the stratospheric Van Allen Belt over Christmas Island three thousand miles away.

We had been rattled, first by a humanly fashioned technological monstrosity; then by an inability to cope with a perfectly natural storm. Fearful hand-of-man nuclearism, and then frightening nature, had exposed our vulnerability — foretelling the dramatic techno-adjustments and climatic-adaptations now confronting humankind.

## OCEANIC WARNING VIII
## NAURU: ECONOMIC COLLAPSE

Historically the people on the remote Central Pacific island of Nauru were free of the likes of the Easter Islander's resource damaging cult. Captain John Fearn the first (1798) European visitor came across a happy and healthy population. He chose the name 'Pleasant Island'.

Fearn's discovery led to another find — a place capped with phosphate, deposited as sea bird guano over millennia. During the middle of the last century almost all that phosphate was stripped out and spread, as fertiliser, onto Australasian and North American farmland; leaving behind a harsh hot moonscape.

The royalty accumulations on behalf of the Nauruans were a minor part of the actual worth of the phosphate; although over time those royalties accumulated as a massive amount of money. This was splurged on an airline, property in Melbourne, a symbolically disastrous opera production, full-on consumerism, and exposure to financial quackery. The Nauruan community soon lost its way within the international money maze.

By the mid-1990s the coffers were empty. The marooned population back on Narau was left financially unraveled. The thirty-year courtship with consumerism had also degraded the people's health and led to a withering of essential fish-catching, food-cropping and coconut-climbing skills. They suffered a downward shift from superrich to barrel-scraping-poor; thirty years in which they were not able to find a new direction, install technological alternatives, establish reality economics, or re-establish their shelter and food supply skills.

This exposure to the global financial contrivance left the community broke and demoralised. Carry forward this global inference: the Nauruan economic lock-down in the wake of a messy fiscal collapse foretelling global lock-down.

# APPENDIX C:
# EXEMPLARS FROM OCEANIA

## EXEMPLAR I
## NIUE: BENIGN GOVERNANCE

Development work took me to the remote Pacific Island state of Niue *(The Rock)* on several occasions spanning three decades.

On my first working day, mid-August 1979, I called on the Administrative Secretary while my wife went off to see the inter-village netball championships. The netball over, a pleasant elderly couple offered her a lift back to the hotel, and during the ride it became apparent that they knew who she was, and all about my whereabouts that afternoon!

At the hotel my wife invited them in for some tea, and to meet me; however I was still with the Administrative Secretary. When they stirred to depart, my wife asked for their names. Straight faced the man said "Tell Robert you've spent the afternoon with the Premier". And, said Mrs Rex "We'll see you tomorrow night for dinner; my husband, who's also Robert, is going fishing first thing in the morning".

Smallness of population, village-sized communities, nobody in the Jail House, self sufficient fishing and agriculture, a sparkling clean environment, observance of traditional fisheries and soil conservation beliefs and practices; in the 1970s these all combined with openly co-operative policy-making to achieve socio-environmental harmony.

# EXEMPLAR II
# NATONGANDRAVUA: LESSON FROM EXPERIENCE

The Fijian village known as Natongandravu lies on the right bank of the considerable Rewa River a few miles inland from its delta. During WWII in the Pacific the canny villagers saved enough from their work on the nearby Nausori Airport to re-site their village and build a collection of valedelakava — houses with corrugated iron walls and roofs. They were all put up in 1946; and by 1966 they were all falling apart!

Over a two-year period I worked on-and-off with the local administrator, Rory Scott, to design a new village layout and organise a rebuilding of the entire settlement as a thirty-unit project. The full story is lengthy; part social engineering, part development experiment, part do-goodery. There was a fairy tale ending when all the replacement houses had been erected; after which Rory and I were ceremonially presented with tambua (giftings to mark happenings of Fijian significance).

For economy and healthiness the houses were clad with plaited bamboo matting sourced and rafted down from a grove twenty miles upstream. This sheathing was designed to blow away in a hurricane; so there would be a constant demand for replacement bamboo. We decided to source some already rooting stock from upriver and raise it locally for future supply.

The first dry season nursery planting's failed. The follow-up wet season planting's also failed. A Sri Lankan bamboo specialist happened to be visiting Fiji. He advised that the wall cladding specie was sensitive to location, unlikely to flourish in an eco-environment, which did not suit exactly. We didn't have the common sense to seek out this knowledge early-on in the project.

# EXEMPLAR III
# REWAUSAU: CIVILISED & SUSTAINABLE

Way back, prior to the Pacific nation's independence, I trekked west-east across Viti Levu, Fiji's largest island — a landmass about the size of Massachusetts. I had the company of Sami as guide, a roll of five-shilling notes, and sevusevu (gifts) for the village stopovers. Our roadhead-to-roadhead trip took seven days; with every bivouac and village sojourn well remembered.

Arriving to stay at Rewausau, pretty much in the middle of Viti Levu, was an education in the harmony of non-fiscal, pro-sustainable, population-balanced living. Sami and I descended on the village from the uninhabitable Rairaimatuku Plateau. About half an hour out we encountered a group of women working their gardens — one fleeing in fright. We dawdled on with the other ladies, which allowed word of our arrival to travel ahead of us, for a house (bure) to be evacuated and readied, and greeting arrangements made.

The house given over to us was basic and spotless, a natural bathing pool was given over as our bathroom; and when we were ready we attended a meke (song and dance) with speeches, food, singing and dancing. Sami and I danced with every woman and child in the village; after which I retired, and the meke continued.

I woke at dawn; above the swishing sound of the nearby creek — singing! Breakfast was brought, after which I returned to the meke, still going strong. Everybody was healthy, well fed, and happy. Money was not evident.

The villagers had basic modern era trappings — machetes, cloth, pots, hand tools and the like — and with these few accessories lived ever-continuing (sustainable) lives of relative contentment; neither degraded nor overcrowded, with the growth of human numbers under control as a consequence of the spacing of pregnancies. Contentment with their lot shone through the well-fed, good-health, cash-devoid and stimulant-free community of Rewausau.

# EXEMPLAR IV
# PAPUA NEW GUINEA: EVERLASTING SUSTAINABLE

Late 1970s: while preparing an environmental impact assessment for a recently roaded district in the Highlands of Papua New Guinea attention was drawn to an untracked area — a tapu (taboo) forest.

In this particular woodland, decades before, the local people had placed their ancestor's on death; then left them reverently alone for the remains to be recycled by nature, with that patch of forest declared off-limits to all. At the time of my visit it was allowable to enter an adjacent part of the tapu forest with a hunting party. Here the deceased were long gone, their assimilation complemented by maturing trees.

By the 1990s those trees were being logged and milled. A macabre, to some, form of Waste Conservancy (ancestors placed decades previously at the base of saplings for natural recycling) complemented by Resource Conservancy (well nourished trees felled decades later to produce construction material). The longer-term sustainable relationship was clear — a parable of pantheistic resilience.

During the course of the 1990 audit I encountered a socio-economic research team from ANU Canberra. Revealing the synthetic status of money; they told the story of a thoughtful response from the leader of one of their research villages in the PNG Highlands. When asked what infrastructure was needed in his district, the chief came up, overnight, with "A bully beef factory, a beer factory, and a money factory."

A facile response? To my way of thinking it is a response pretty much in line with Western delusional expectations about faux wealth and excess consumption.

# EXEMPLAR V
# COOK ISLANDS: TECHNOLOGICAL SUFFICIENCY

Early 1980s. During the preparation of a Country Report for the Cook Islands I spent a few days working on Aututaki; curiously, another hillocked atoll like Malima (Oceanic Warning VII). 'As one does' in Oceania on a Sunday my wife and I were at church, politely attentive to the service and the harmonies sung in Maori, absorbed by the view out across the pale green shallows to the dark blue rollers curling, smashing and booming into brilliant white foam on the fringing coral reef.

After church, we set off on our hired scooter. With the tide going out, the plan was to travel to the uninhabited southern tip of the atoll, Lands End. We had shop food — ships biscuits (Fiji), canned tuna (Canada), coke (Australia) — in the scooter's pannier. For quite a long way, maybe two miles, our course followed the lonely shore around point after point.

Rounding the tip of the island at Lands End we came upon a family.

The father and his young son were wading ashore with fish they had speared. A young girl was tending a fire and boiling a pot of yams. Mother was sitting cross-legged in the shade nursing her youngest. A tethered horse, straight from Gauguin, was grazing nearby. We chatted. The man cleaned the fish and put them over the embers, the boy climbed a coconut tree to knock down some green drinking nuts. There was much talking and fussing. We produced our shop food. Grace given; we tucked in. The children fell on the shop-food delights; we revelled in the excellence of just-caught grilled fish, boiled yam and the drinking coconuts.

After a rest we returned to the settlement along an inland route; kids on the back of the scooter, my wife and the baby riding the horse, the man leading. Our hosts were seen by us to be both happy and poor — twenty remittance dollars a week poor — their lives had a harsh edge.

We enjoyed our time together. The lunch our hosts provided was gourmet; our shop-food and soda drink contribution unconscionable; their carbon footprint negligible, ours execrable. Without doubt they wished for lives richer in technological texture, with more of the complex baubles we had in our lives. Yet we, satiated with comfort travel and goods, envied the simple satisfactions they shared with us; each yearning for more of what was in the lives of the other.

# EXEMPLAR VI
# MELANESIA-POLYNESIA: PRAGMATIC ADEQUACY

Fiji's medical service began in my time there with the training of Assistant Medical Officers; people to whom by-pass surgery would be a mystery. These were practitioners well able to get a baby or an abscessed tooth out with ease; dispensing sound, low-tech, low-cost, proficient medical attention to remote villagers, and indeed on occasion to myself and my companions, in the wilderness parts of PNG, the Solomons and, of course, Fiji.

Samoan hotels range from the four star Aggie Grey, to $20 a night right-on-the-beach fales (palm leafed sleeping platforms with a mattress and mosquito net, and a nearby community washroom and store) — Idyllic.

On Rarotonga where gasoline has always been expensive, vehicles move at a stately twenty miles an hour. In the past when tyres eventually wore out they were dumped, and new imports fitted. But retreaded tyres — regarded in high speed industrial societies as unsafe and liable to flail — work well on low-speed Rarotonga. In a brilliant move a retread station was set up; and in a blink imports were slashed. I asked the station operator whether a tyre could be retreaded more than once? "Yeah Mate, sweet as; up to four times" was his laconic reply.

# APPENDIX D:
# MONETARY STABILITY

At the end of this writing I am struck by the audacity of the stable money, stable society, stable biosphere polemic; about which there is no doubt in terms of necessity, although there are serious misgivings in terms of realisation. This Appendix furthers that audacity with a take on two of our fiscal system's most serious failings, sovereign reserve stability, and debt obligation servicing.

The underlying need is for politicians to bolster their economic understanding and responsibility around banking, speculation, lending and debt; making policy and taking decisions on fiscal resilience and debt avoidance as a political function, even when they do not have a background in money management.

Global Reserve, and Debt Obligation issues cannot be left to bankers, hedge fund fixers, insurance re-investors and the like. Politicians have to take responsibility to guide their economies back into budgetary surplus on behalf of those who entrusted them with their votes to do this job for them; and, important this, in intergenerational fairness to our children.

Global Reserve Resilience and Stability, and Debt Obligation Auditing, when linked with balanced in-country budgeting, would divert this generation from the theft of our children's future, and lessen the spoliation and destruction of the habitat our offspring will inherit.

## GLOBAL RESERVE RESILIENCE & STABILITY

Quite how the United States dollar holdings, raised trillion-by-trillion year-on-year to cover debt obligations and pay running expenses, maintains validity as a global benchmark currency, beats me.

A benchmark currency needs to be, surely, near inflation-proof, of near-fixed amount and rock-solid — graded AAA perpetually. Fiscal equilibrium is of paramount global importance because it can ensure social stability and underpin environmental wholesomeness. An important angle is that most short-term politicians do not appreciate that, for significant longer-term reasons, annual fiscal outgoing must not exceed annual income. The role of politicians, as statespeople, is to ensure that putting a halt to future debt-burdening and habitat-destruction are goals-in-common; whereas in political fact they have proved politically too awkward, long-term and complex for them to enforce.

The perspective from China, as virtual comptroller of the Mighty Dollar, is that US 'sovereignty' is now fraught by a stateside inability to wean the nation off the regular money supply increases that unsettle every exchequer and pocketbook throughout the globe.

Little wonder that China now identifies the need for a New Global Reserve Currency; something like a return to the former Gold Standard, without the gold. The role for such a currency being to ensure the reliability of country-to-country transactions and payments.

## DEBT OBLIGATION AUDITING

Nations on the EU periphery, poor nations suckered into accepting loans to install military infrastructure, Settler Societies both printing and borrowing money to feed the consumer maw, and all those economically colonised by corporations and cartels, need to face up to auditing their debt burdens: — identifying 'Legitimate Debt' for eventual full repayment, 'Pernicious Debt' for partial repayment, and 'Toxic Debt' for non-payment.

Debt Obligation Auditing can and should be authorised through referenda; carried forward by an electoral commission. Analysts would put up background papers on how economic affairs got to where they are, an all-party selection of currently elected politicians assigned for procedural overview, and actioned by a core Debt Audit Committee elected from the broad church of life. It would be that Committee's job to conduct the enquiry, report findings and make recommendations to their government.

Better to bite the bullet before the likes of the fate which has befallen Iceland, Ireland, Greece; and Thailand, Indonesia, Korea; and Chile, Argentina, Ecuador; and Egypt, Libya, Syria; breaks over other homelands. Public Debt must be dissected, analysed, understood; and actioned to create conditions for ensuring a sustainable socio-economic future.

# APPENDIX E:
# PART B: SUMMARY

| | | | | |
|---|---|---|---|---|
| **9** | **New Paradigm: Norms: Politics** | | **10** | **Information and Education** |

**9  New Paradigm: Norms: Politics**

- Foretell risks, uncertainties, limits
- Sustainability: a creative human right
- Act morally, decently, assertively
- Fashion left-right institutionalism
- Converge: politics, science, economics
- Maintain a CommonGood mindset
- Sue for environmental justice
- Localise decision-taking

**10  Information and Education**

- Heed historical precedents
- Identify and respect planetary resilience
- Learn from the four-fifths poor
- Harness electronic capabilities
- GobalGood data, analysis, prognosis
- Sustainable culture and pedagogy
- Get wise, think sharp, slow down

**11  Demographics**

- Empower women
- Alleviate fear of old age isolation
- Educate on reproductive health
- Provide reproductive services Leadership: political and religious

**Economics**

- Establish a stable global reserve
- Institute non-speculative systems
- Blitz fiscal scandals and scams
- Prohibit off-balance sheet activities
- Hobble options and futures trading
- Install 'transaction' taxes
- Poleaxe harmful to climate subsidies

**12  Resource with Waste Conservancy**

- Contraction of inefficiencies
- Convergence of efficiencies
- Empower environmental defence
- Think global: Act local
- Live lightly lastingly
- Small scale, ecologically sustainable
- Waste: reuse recycle retrofit
- Install input-output modelling

**13  Goals to Fulfil**

- Respect resilience boundaries
- Follow 'soft' pathways 'dematerialise'
- Conserve finite irreplaceable resources
- Capture freeflow energy
- Punish environmental criminality
- Seek out environmental justice
- Establish a GlobalGood bureau

  (The 'Goals' chapter lists 45 Protocols: National, Community and Local)

**14  Targets**

- Softer cleaner waste-avoiding
- 'Less' where it achieves 'More'
- Forest protection and expansion
- Biospheric purity
- Sourcing-fabrication-consumption
- Freeflow energy capture
- Natural climate control
- Eco-design and GreenTick practices
- Transport Demand Management

# REFERENCES

References dated prior to 2000, The Millennium Date, tend to reveal analyses and concerns. Those after the Millennium Date trend toward prognoses and answers.

Adelson, J. *Inventing Adolescence*, Transaction Books, 1986.
Arendt, R. *Growing Greener*, Island Press, 1999.
Ayres, R.U. and E.H. *Crossing the Energy Divide,* Wharton School Publishing, 2009.
Barber, R.B. *Consumed: How Markets Corrupt Children and Infantalize Adults*, Norton, 2007.
Barnett, T.P.M. *A Blueprint for Action; A Future Worth Creating,* Putnam, 2005.
Barnham, K. *The Burning Question*, Weidenfeld & Nicholson, 2014.
Benes J. and M. Kumhof (IMF Working Paper) *The Chicago Plan Revisited*, IMF Press, 2012.
Benkler, Y. *The Wealth of Networks*, Yale University Press, 2006.
Bernhardt, J. *A Deeper Shade of Green,* Balasoglou Books, 2008.
Berger, P.L. *Adventures of an Accidental Sociologist*, Prometheus, 2011.
Berry, W. *What Are People For?*, North Point, 1990.
Blainey, G. *Triumphs of the Nomads: A History of Ancient Australia,* MacMillan Press, 1975.
Blaxter, K. *People, Food and Resources*, Cambridge University Press, 1989.
Bok, D. *The Politics of Happiness,* Princeton University Press, 2010.
Borstein, D. *How to Change The World: Social Entrepreneurs and the Power of New Ideas*, OUP, 2004.
Bosselman, K. *When Two Worlds Collide*, RSVP Press, 1995.
Bower, T. *The Squeeze: Oil Money and Greed* Harper Press, 2009.
Brown, L.R. *Beyond Malthus*, Norton, 1999.
Brown, L.R. *Plan B 2.0: Rescuing a Planet Under Stress and a Civilization in Trouble*, Norton, 2006.
Brown, L.R. *Eco-Economy*, Norton, 2001.
Brundtland, H. *Our Common Future*, OUP, 1987.
Bryson, B. (editor) *Seeing Further: The Story of Science and the Royal Society,* Harper Press, 2010.

Cable, V. *The Storm: The World Economic Crisis and What it Means*, Atlantic, 2009.

Camejo, P. *The Socially Responsible Investing Advantage*, New Society Publishers, 2002.

Capra, F. 'The Turning Point' *Science, Society and the Rising Culture*, Berkeley, 1975.

Carbon Tracker Initiative. *Unburnable Carbon 2013*, (With Grantham Research Institute) 2013.

Carson, R. *Silent Spring*, Houghton Mifflin, 1962.

Chan, S. *The End of Certainty*, Zed Books, 2010.

Chang, H-J. *23 Things They Don't Tell You About Capitalism*, Allen Lane, 2010.

Charter, M. with Ursula Tischner. *Sustainable Solutions: Products and Services*, Greenleaf, 2001.

Christensen, C. *The Innovator's Dilemma*, Harvard Bussiness School Press, 1997.

Crumlish, C. *The Power of Many*, Sybex, 2004.

Collier, P. *The Bottom Billion: Why the Poorest Countries are Failing*, Bodley Head, 2006.

Collin, R.M. and R.W.Collin. *Encyclopedia Of Sustainability*, Greenwood Press (3 volumes), 2010.

Daly, H.E. with J.B.Cobb. *For the Common Good*, Beacon, 1989.

Daniels, T. *When City and Country Collide*, Island Press, 1999.

Datcschefski, E. *The Total Beauty of Sustainable Products*, Rotovision, 2001.

Dauncey, G. with Ross Gelbspan. *Stormy Weather: 101 Solutions to Climate Change*, New Society, 2001.

Dawkins, R. *The Ancestor's Tale; A Pilgrimage to the Dawn of Life*, Weidenfeld and Nicholson, 2004.

Dawkins, R. *The Greatest Show on Earth*, Bantam Press, 2009.

Desai, P. with Sue Riddlestone. *BioRegional Solutions for Living on One Planet*, Green Books, 2003.

Diamond, J. *Collapse: How Societies Chose to Fail or Succeed*, Penguin, 2005.

Dilworth, C. *Too Smart for Our Own Good*, Cambridge University Press, 2009.

Dorfman, J. *The Lazy Environmentalist*, Stewart Tabori Chang, 2007.

Dworkin, R. *Justice for Hedgehogs*, Belknap Press, 2010.

Eagleton, T.E. *The Meaning of Life*, OUP, 2007.

Ehrenreich, B. *Bait and Switch: Futile Pursuit of the American Dream*, Metropolitan, 2005.

Ehrenreich, B. *Fear of Falling: The Inner Life of the Middle Classes*, Harper Perennial, 1990.

Ehrlich, P. with A. H. Ehrlich and J. P. Holdren. *Ecoscience*, Freeman Press, 1977.

Elgin, D. *Awakening Earth: Exploring the Evolution of Human Culture*, William Morrow, 1993.

Escobar, A. *Encountering Development*, Princeton University Press, 1995.

Estill, L. *Biodiesel Power*, New Society Publishers, 2005.

Esty, D. with A.Winston. *Green to Gold*, Yale University Press, 2006.

Etzioni A. *The Spirit of Community*, Gown Publishers, 1993.

Fahrenthold, D.A. 'Holding tight to the status quo' *Washington Post*, January 2010.

## REFERENCES

Faud-Luke, A. *Eco-Design: The Sourcebook*, Chronicle Books, 2002.

Flannery, T. *Here on Earth: A New Beginning*, Penquin, 2011.

Flannery, T. *The Future Eaters*, Reed Books, 1994.

Flannery, T. *The Weather Makers*, Penquin, 2007.

Florida, R. *The Rise of the Creative Class*, Basic Books, 2004.

Fodor, E. *Better Not Bigger,* New Society Press, 1999.

Foster, N. *Reflections*, Prestel Publications, 2005.

Foucault, M. *The Order of Things,* Random House, 1992.

Franzen, J. *Freedom,* Farrar Straus and Giroux, 2010.

Frost, Robert. *Selected Poems,* Gramercy Books, 2001.

Friedman, T. *Hot, Flat and Crowded,* Farrar, Straus and Giroux, 2008.

Fukuyama, F. *The End of History and the Last Man*, Free Press, 1992.

Galbraith, J.K. *The Culture of Contentment*, Houghton Mifflin, 1992.

Gaynor, B. 'Twin Towers of Debt' *NZ Herald*, September 27 2008.

Garreau, J. *Edge City: Life on the New Frontier*, Anchor Press, 1992.

Garreau, W.M. *Radical Evolution*, Doubleday, 2004.

Geddes, P. *Cities in Evolution*, Williams and Norgate, 1915.

Gee, M. 'Beyond Ending: Looking into the Void' in *Seeing Further,* (edited by Bill Bryson) Harper, 2010.

Gelbspan, R. *The Heat Is On: The Climate Crisis, The Cover-up*, Perseus Books, 1998.

Georgescu-Rogen, N. *The Entropy Law*, Harvard University Press, 1971.

Giddens, A. *The Politics of Climate Change*, Polity Press, 2009.

Gilpin, A. *Dictionary of Economic Terms*, Butterworths, 1973 (Third Edition).

Gipe, P. *Wind Power: Renewable Energy for Home Farm and Business*, Chelsea Green, 2004.

Glacken, C.J. *Traces on the Rhodian Shore,* University of California Press, 1977.

Gladwell, M. *The Tipping Point,* Little Brown, 2002.

Glover, J. *Humanity: A Moral History of the 20th Century*, Jonathan Cape, 1999.

Goldsmith, E. and others. *A Blueprint for Survival*, Penguin, 1972.

Goldsmith, E. *The Way: An Ecological World-View*, Rider, 1992.

Goldsmith, Z. *The Constant Economy*, Atlantic Books, 2009.

Goodall, C. *How to Live a Low-Carbon Life*, Earthscan, 2007.

Goodall, C. *Ten Technologies to Save the World*, GreenProfile, 2008.

Goodstein, D. *Out Of Gas*, Norton, 2004.

Gore, Al. *An Inconvenient Truth*, Rodale Press, 2006.

Gore, Al. *Earth in the Balance: Ecology and the Human Spirit,* Plume Press, 1994.

Gore, Al. *Our Choice: A Plan to Solve the Climate Crisis,* Bloomsbury Publishing, 2009.

Goudie, A. *Human Impact on the Natural Environmentt*, Blackwell, 1999.

Gray, J. *Gray's Anatomy: Selected Writings*, Allen Lane, 2009.

Gray, J. *The Silence of Animals: On Progress and Other Modern Myths*, Penguin, 2013.

Green, D. *From Poverty to Power*, Oxfam International, 2008.

Green, J.M. *Deep Democracy*, Rowman and Littlefield, 1999.

Greenspan, A. 'We need a better cushion against risk' *Financial Times*, 26 March 2009.

Greider, W. *One World, Ready or Not*, Simon and Schuster, 1997.

Haar, C.M. *The End of Innocence*, Scott Foreman, Illinois, 1972.

Hamilton, C. *Requiem for a Species*, Earthscan, 2010.

Hansen, J. *Storms of My Grandchildren*, Bloomsbury, 2009.

Hardin, G. *Living Within Limits*, OUP, 1993.

Hardin, G. 'The Tragedy of the Commons', *Science*, No. 162, 1968.

Hart, S.L. *Capitalism at the Crossroads; Unlimited Business Opportunities*, Wharton Publisers, 2005.

Harre, N. *Psychology for a Better World*, Self published, 2011.

Havel, V. *The Art of the Impossible: Politics and Morality in Practice*, Knopf, 1997.

Harvey, D. *The Enigma of Capital: And the Crises of Capitalism*, OUP, 2010.

Hawken, P. with Amory Lovins and Hunter Lovins. *Natural Capitalism*, Earthscan, 1999.

Hawking, S. *A Brief History of Time*, Bantam, 1989.

Heilbroner, R. *Business Civilization in Decline*, Norton, 1976.

Heller, J. *Catch-22*, Simon and Schuster, 1961.

Hensher, D. A. with K.J. Button *Handbook of Transport and the Environment*, Elsevier, 2003.

Henson, R. *Rough Guide to Climate Change*, Rough Guides Ltd, 2006.

Hobsbawn, E.J. *The Age of Capital 1848–1875*, Weidenfeld and Nicholson, 1975.

Holister, G. and Andrew Porteous *The Environment: A Dictionary of the World Around Us*, Arrow, 1976.

Holmgren, D. *Permaculture: Principles and Pathways Beyond Sustainability*. Holmgren Studios, 2002.

Homer-Dixon, T. *The Ingenuity Gap*, Knopf, 2000.

Hopkins, R. *The Transition Handbook*, Green Books, 2008.

Illich, I. *Tools for Conviviality*, Fontana-Collins, 1973.

IPCC (International Panel on Climate Change) *Fourth Assessment Report*, UNFCC/WMO/UNEP, 2007.

Jaccard, M. with John Nyboer and Bryn Sadownik. *The Cost of Climate Policy*, Eurospan, 2002.

Jackson, T. *Prosperity Without Growth: Economics for a Finite Planet*, Earthscan, 2010.

James, O. *Affluenza*, Vermilion Press, 2007.

Jesson, B. *Only Their Purpose is Mad*, Dunmore Press, 1999.

Judt, T. *Ill Fares the Land*, Penguin Books, 2010.

Juniper, T. *What Has Nature Ever Done For Us?* Profile Press, 2013.

Kahane, A. *Solving Tough Problems*, Berrett-Koehler, 2004.

Kamenetz, A. *Generation Debt: How Our Future Was Sold Out*, 2006.

Kaplan, R.D. *An Empire Wilderness*, Random House, 1998.

Kelly, E. *Powerful Times: Rising to the Challenge of Our Uncertain World*, Wharton Publishing, 2006.

Kelly, K. *Out of Control,* Addison-Wesley, 1994.

Kemp, W.H. *Smart Power: An Urban Guide to Renewable Energy,* Aztext Press, 2005.

Keynes, J.M. *The Collective Writings,* Macmillan, 1971.

Kingdon, J. *Agendas, Alternatives and Public Policies,* Longman, 1995.

Klein, N. *No Logo,* Picador, 2002.

Klein, N. *The Shock Doctrine,* Metropolitan Books, 2008.

Koenig, K.W. *The Rainwater Technology Handbook,* Wilo-Brain, 2001.

Kohr, L. *The Breakdown of Nations,* Routledge and Kegan Paul, 1957.

Korten, D.C. *The Post-Corporate World,* Berrett-Koeehler, 2000.

Korten, D.C. *When Corporations Rule the World,* Earthscan, 1995.

Kropotkin, P. *Fields Factories and Workshops Tomorrow,* Freedom Press, 1985 (first published 1889).

Kuhn, T.S. *The Structure of Scientific Revolutions,* University of Chicago Press, 2005.

Kunstler, J.H. *The Geography of Nowhere,* Simon and Schuster, 1993.

Kunstler, J.H. *The Long Emergency,* Atlantic Books, 2005.

Lanchester, J. *Whoops! Why Everyone Owes Everyone and No One Can Pay,* Allen Lane, 2010.

Langdon, P.A. *A Better Place to Live,* 1995.

Leakey, R. with Roger Lewin. *The Sixth Extinction,* Anchor Books, 1996.

Leslie, J. *The End of the World,* Routledge, 1996.

Lewis, H. with John Gertsakis. *Design and Environment,* Greenleaf, 2001.

Lomberg, B. *Cool It: The Skeptical Environmentalist's Guide,* Marshall Cavendish, 2007.

Lomberg, B. *Smart Solutions to Climate Change,* Cambridge University Press, 2010.

Lomberg, B. *The Skeptical Environmentalist,* Cambridge University Press, 2001.

Lovelock, J.E. *The Revenge of Gaia,* Allen Lane, 2006.

Lovelock, J.E. *The Vanishing Face of Gaia,* Allen Lane, 2009.

Lovins, B. *Soft Energy Paths,* New York, Ballinger, 1977.

Lowe, S. and Alan McCarthur. *Is It Just Me Or Has the Shit Hit the Fan,* Little Brown, 2009.

Lynas, M. *High Tide: The Truth About Our Climate Crisis,* Picador, 2004.

Lynas, M. *The God Species: How the Planet Can Survive the Age of Humans,* Fourth Estate, 2011.

Lynas, M. *Six Degrees: Our Future on a Hotter Planet,* Fourth Estate, 2007.

Maass, P. *Crude World: The Violent Twighlight of Oil,* Allen Lane, 2009.

Mackay, D. *Sustainable Energy Without the Hot Air,* www.withouthotair.com, 2009.

MacGregor, N. *A History of the World in 100 Objects,* Allen Lane, 2010.

Malthus, T.R. *An Essay on the Principle of Population,* Everyman Press, 1914 (First published 1798).

Mann, M.E. *The Hockey Stick and the Climate Wars,* Columbia University Press, 2012.

Marx, K. *Capital: Volume I,* Penguin, 1976. (First published 1867).

Maslow, A.H. *Toward a Psychology of Being,* Reinhold, 1968.

Maskrecki, P. *The Smaller Majority*, Belknap Press, 2005.

Matheson, C. *Green Chic: Saving the Earth in Style*, Sourcebooks, 2008.

Mau, B. *Massive Change*, Phaidon Press, 2004.

McCamant, K. and C. Durrett. *CoHousing: A Contemporary Approach to Housing Ourselves*, Habitat, 1990.

McCarthy, C. *The Road*, Knopf, 2006.

McDonough, W. with Michael Braungart. *Cradle to Cradle*, North Point Press, 2002.

McHarg, I.L. *Design with Nature*, Doubleday, 1971.

McKibben, B. *The End of Nature*, Random House, 1989.

McKibben, B. *Maybe One*, Simon and Schuster, 1998.

McKinney, M,L., with R.M.Schoch. *Environmental Science*, Jones and Bartlett, 1998.

McNeil, B. *The Clean Industrial Revolution*, Allen and Unwin, 2009.

McNeill, J.R. *Something New Under The Sun: Environmental History of the 20th Century*, Norton, 2000.

Meadows, D.H. with D. L. Meadows, J. Randers and W. Behrens. *The Limits to Growth*, Earth Island, 1972.

Meadows, D.H. with Meadows and Randers. *Limits to Growth: 30-year Update*, Chelsea Green, 2004.

Mill, J.S. *Utilitarianism* (Edited by Mary Warnock) New American Library Press, 1974. (Original 1859).

Mollison, B. *Permaculture: A Designers' Manual*, Tyalgum Press, Australia, 1988.

Monbiot, G. *Heat: How To Stop the Planet Burning*, Allen Lane, 2006.

Mumford, L. *Technics and Civilization*, New York, Harcourt and Brace 1934.

Munn, T. (Editor) *Encyclopaedia of Global Environmental Change*. Wiley (5 volumes) 1992.

Murphy, T.W. 'Beyond Fossil Fuels' in *State of the World 2013*, Island Press, 2013.

Nair, C. *Consumptionomics: Asia's Role in Reshaping Capitalism*, Wiley, 2011.

Nattrass, B. with M. Altomare. *Dancing with the Tiger: Sustainability Step by Step*, New Society, 2002.

Norgaard, R.B. *Development Betrayed: The End of Progress,* Routledge, 1994.

Nozick, R. *Anarchy, State and Utopia*, Basic Books, 1974.

Nye, J.S. *Soft Power*, Public Affairs Press, 2005.

Oreskes, N. with Erik Conway. *Merchants of Doubt*, Bloomsbury, 2010.

Orr, D. *Ecological Literacy*, University of New York Press, 1992.

Osborn, A. *Applied Imagination,* Scribner, 1954.

Owen, D. *Green Metropolis: Living Smaller and Living Closer Are Keys to Sustainability,* Riverhead, 2009.

Owens, S. *Energy, Planning and Urban Form,* Pion, 1986.

Papanek, V. *The Green Imperative*, Thames and Hudson, 1995.

Patel, R. *The Value of Nothing: How to Reshape Market Society and Redefine Democracy*, Portobello, 2010.

Perry, R.B. *Realms of Value*, Harvard University Press, 1954.

Plimer, I. *Heaven and Earth: Global Warming: The Missing Science*, 2009.

Plumwood, V. *Feminism and the Mastery of Nature*, Routledge, 1993.

Pollan, M. *The Omnivor's Dilemma*, Penguin Press, 2006.

Ponting, C. *A Green History of the World,* Sinclair-Stevenson, 1991.

Popper, K. *The Open Society and Its Enemies*, 2 Vols: Routledge & Kegan Paul, 1974. (Original 1945).

Porrit, J. *Capitalism as if the World Mattered,* Earthscan, 2005.

Price, D. *Radical Simplicity: Creating an Authentic Life*, Running Press, 2005.

Putnam, R. D. *Bowling Alone*, Simon and Schuster, 2000.

Ramage, J. *Energy: A Guidebook*, Oxford University Press, 1997.

Rand, A. *For the New Intellectuals: The Philosophy of Ayn Rand*, Signet, 1996.

Rand, A. *We the Living*, Signet, 1996.

Rawls, J. *Theory of Justice*, Cambridge, Mass., Harvard University Press, 1971.

Rees, M. *Our Final Century?* Heinemann, 2004. [Published in the United States as *Our Final Hour.* 2001.]

Restakis, J. *Humanizing the Economy: Co-operatives in the Age of Capitalism*, New Society Press, 2010.

Rexroth, K. *Communalism: From Its Origins to the Twentieth Century*, New York, Seabury Press, 1974.

Riddell, R. B. *Ecodevelopment: Economics Ecology and Development*, Gower and St Martins, 1981.

Riddell, R. B. *Regional Development Policy*, Gower (London) and St Martins (New York), 1985.

Riddell, R. B. *Sustainable Urban Planning: Tipping the Balance*, Blackwell-Wiley, 2004/07.

Rifkin, J. *The Hydrogen Economy*, Tarcher, 2003.

Rittel, H. and M.W. Webber. 'Dilemmas in a General Theory' in *Policy Sciences,* 4, 1973.

Robert, K-H. *The Natural Step Story*, Gabriola Island Press, 2002.

Roberts P. *The End of Oil*, Houghton Mifflin, 2004.

Roberts T. with B. Parks. *A Climate Of Injustice*, MIT Press, 2007.

Robertson, J. *The Sane Alternative*, Villiers Publications, 1978.

Robinson, J. with Francis, Legge and Lerner. 'Defining Sustainable Society', *Alternatives,* Vol. 17, 1990.

Rockstrom, J. with 26 others. 'A Safe Operating Space for Humanity', *Nature,* 461 September 2009.

Rogers, H. *Gone Tomorrow: The Hidden Life of Garbage*, New Press, 2005.

Rogers, H. *Green Gone Wrong*, Scribner, 2010.

Rosebraugh, C.S. *Greedy Lying Bastards* [film], One Earth Productions, 2013.

Rousseau, J.J. *The Social Contract*, Dent, 1973. (First published 1762).

Roy, A. *Broken Republic: Three Essays*, Hamish Hamilton, 2011.

Sachs, J.D. *Common Wealth: Economics for a Crowded Planet*, Allen Lane, 2008.

Sahlin, M. *Stone Age Economics*, Aldine-Atherton, 1972.

Sandel M.J. *Justice: A Reader*, Allen Lane, 2009.

Sandel M.J. *Justice: What's the Right Thing to Do?* Allen Lane, 2009.

Sandel M.J. *What Money Can't Buy: The Moral Limits of Markets*, MacMillan, 2013.

Satterthwaite, D. 'Population and Urbanisation' *Environment and Urbanisation,* Vol 21 No 2, 2009.

Satterthwaite, D. and others. *Environment Problems in an Urbanising World,* Earthscan, 2009 edition.

Saul, J.R. *On Equilibrium,* Penguin, 2002.

Schaeffer, J. *Solar Living Sourcebook,* New Society Publishers, 2005.

Schama, S. *Landscape and Memory,* Knopf, New York, 1995.

Schulz, K. *Being Wrong,* Portobello, 2010.

Schumacher, E.G. *Small is Beautiful: A Study of Economics as if People Mattered,* Abacus, 1974.

Schumpeter, J. *History of Economic Analysis,* OUP, 1954.

Schneider, S.H. 'Confidence Consensus and Uncertainty...' *Seeing Further,* (Ed. Bill Bryson) Harper, 2010.

Seidl, A. *Finding Higher Ground: Adaptation in the Age of Warming,* Beacon Press, 2011.

Sen A. *The Idea of Justice,* Penguin-Allen Lane, 2009.

Sennett, R. *The Corrosion of Character: Personal Consequences of Work in New Capitalism,* Norton, 1998.

Sennett, R. *The Craftsman,* Yale University Press, 2008.

Sennett, R. *Together: Rituals Pleasures and Politics of Co-operation,* Yale University Press, 2012.

Sharp, G. *Waging Nonviolent Struggle: 20th Century Practice and 21st Century Potential,* Horizon, 2005.

Simons, D. with C. Chabris *The Invisible Gorilla,* Crown Publishing, 2010.

Smith, A. *An Inquiry Into the Nature and Causes of the Wealth of Nations,* Dent, 1910. (Original 1776).

Smith, R. 'Green capitalism: the God that failed' in *Real World Economics Review,* March 2011.

Socolov, R.H. with S.W. Pacala 'A Plan to Keep Carbon in Check' *Scientific American,* September 2006.

Solnit, R. *A Paradise Made in Hell,* Viking, 2009.

Sorkin, A.R. *Too Big To Fail,* Allen Lane, 2009.

Soros, G. 'The Capitalist Threat' in *The Atlantic Monthly,* February 1997.

Steffen, A. (Editor) *Worldchanging,* Abrams, 2006.

Steffen, A. and others. *Global Change and the Earth System: A Planet Under Pressure,* Springer, 2004.

Sterling, B. *Tomorrow Now,* Random House, 2002.

Stern, N. *A Blueprint for a Safer Planet,* Bodley Head, 2009.

Stewart, J.B. *Tangled Webs: How False Statements are Undermining America,* Penguin Press, 2011.

Stiglitz, J. *Freefall,* Allen Lane, 2010.

Stiglitz, J. *Globalisation and its Discontents,* Allen Lane, 2002.

Stiglitz, J. *The Price of Inequality,* Allen Lane, 2012.

Stiglitz, J. 'Beyond Market Fundamentalism', *Annals of Public and Co-operative Economics*, 80:3, 2009.

Sucher, D. *City Comforts*, City Comforts Press, 1995.

Sunstein, C. *On Rumours: How Falsehoods Spread*, Allen Lane, 2009.

Suzuki, D. with Amanda McConnell. *The Sacred Balance*, Greystone Books, 1997.

Suzuki, D. with Holly Dressel. *Good News for a Change*, Greystone Books, 2002.

Swift, R. *Democracy*, New Internationalist 'No-Nonsense' Series, 2009.

Szaz, A. *Shopping Our Way to Safety*, University of Minnesota Press, 2007.

Taylor, N. *The Village in the City,* Maurice Temple Smith, London, 1973.

Thackara, J. *In The Bubble*, MIT Press, 2005.

Thayer, R.L. *Grey World Green Heart,* Wiley, 1994.

Tickell, O. *Kyoto2: How to Manage the Global Greenhouse,* Zed Books, 2008.

Todd, N.J. *A Safe and Sustainable World: The Promise of Ecological Design*, Island Press, 2005.

United Nations *Resilient People, Resilient Planet: A future worth choosing,* 2012.

United Nations (Conference on Environment and Development). *Agenda 21: The Rio Declaration*, 1992.

United Nations Development Programme *Human Development Report 2007/2008,* Palgrave 2007.

Urban Land Institute. *Mixed Use Development Handbook,* Urban Land Institute, Washington, 1987.

Visser, M. *The Way We Are,* Harper Collins, 1994.

Wackernagel, M. and W. Rees. *Our Ecological Footprint,* New Society Publishers, 1996.

Weart, S.R. *The Discovery of Global Warming*, Harvard University Press, 2003.

Weisman, L.K. *Descrimination By Design: A Feminist Critique*, University of Illinois, 1992.

Wilkinson, R. and K. Pickett. *The Spirit Level: Why More Equal Societies Do Better*, Allen Lane, 2009.

Wilson, E.O. *The Future of Life*, Littlejohn, 2002.

Wise, J. *Extreme Fear: The Science Of Your Mind In Danger*, Palgrave Macmillan, 2010.

Wittgenstein, L. *Collected Works* (edited), Blackwell, 1998.

Worldwatch Institute. *State of the World 2013: Is Sustainability Still Possible*, Island Press, 2013.

Womack, J. with D. Jones. *Lean Thinking*, Simon and Schuster, 1996.

Worster, D. *Nature's Economy*, Cambridge University Press, 1977.

Wright, R. *A Short History of Progress*, Carrol and Graf, 2005.

Zencey, E. 'Energy as Master Resource' in *State of the World 2013: Is Sustainability Still Possible*, 2013.

# ENDNOTES

## CHAPTER 1: THREE COMMUNITIES: TWO FAILURES

1. **Here workers were…** Visits to The Grindery Stream left us in awe of the place and a sense of being in touch with our early origins. We took away some of the discarded shards, mostly badly formed adzes which hadn't shaped-up. With one find I was able to locate the matching half, and later 'two-pot' glue them together, and in this way bridge to the artisan's frustration, over 20,000 years ago, at the breakage of his or her beautifully crafted object.

2. **The process cost…** According to a neighbour, Nigel Pricket (archaeologist) the notion of the Easter Island story being 'not settled at the time of our visit (1971)' puts me among the contrarians, for the enigma of Easter Island (more appropriately Rapanui) was, apparently, sorted out before then!

3. **Every human grouping…** One urban place can be more 'sustainable' than another. It is not logical to claim any urban place as sustainable in-and-of itself.

4. **With less than half…** Martin Rees is Britain's Astronomer Royal. His book's title in full is *Our Final Century: Will the Human Race Survive the 21st Century?* (2003).

5. **Relative to longer gone…** My 2004 *Sustainable Urban Planning* included a sidebar 'Easter Island: Earth Island' citing Clive Ponting's *A Green History of the World* (1991). The 'Easter Island as a metaphor for Earth Island' fascination continues, with a deluge of writings, from which a good point of departure is *The Enigmas of Easter Island* (2003) by Paul Bahn and John Flenley.

6. **A common thread…** On my first east-west flight across the United States in the early 1960s I noticed that the rural landscape out of Chicago comprised 'quarter squares' mostly divided into three or four prosperous farmlets. Half an hour later the original 'quarter squares' were intact, another half-hour further on and four of the original 'quarter squares' were amalgamated to form a square mile farm, and yet another half hour further on the landscape (Kansas?) seemed uninhabited with only traces of the original road lines showing through. Also consult Timothy Egan (2006) *The Worst Hard Time: The Story of Those Who Survived the Great American Dust Bowl.*

7  **The presence of humankind**...Ludwig Wittgenstein coined the phrase 'simply a wonderment' as his explanation for the presence and existence of life on earth.

8  **The populations of...** OECD — thirty countries, of above average wealth, centered on Europe and North America with outliers in Japan-Korea and Australasia. Annual operating expenses of the OECD are about equal to the production costs for King Kong the movie! OPEC — thirteen oil producing nations, mostly in the Middle East; with no disclosure of operating expenses.

9  **Never prior to this past century...** 'Disorder' (entropic disorder) is expressive of the way the uptake and use of finite resources (most notably mineral fuels, but also other finite resources) are transformed (disordered dissipated degraded) with use, into wastes of no utility — for example, after combustion carbon fuels dissipate into heat, carbon dioxide and other useless gasses.

10  **One remedy involves...** The WWF *Nature's Living Planet* lists and ranks global hectares per person. A variant descriptor to the footprint concept has been devised by the New Economics Foundation. By Easter, in a notional year, most European nations, along with the USA and Japan, will cease to be self-sufficient; thereafter, throughout the rest of the notional year, they live off other nations.

11  **Simply put...** Genocide: an observation. Hittlerism occurred within virtually land-locked 'civilised' Germany. In rural rusticity, arose the genocide from also land-locked Rwanda, Burundi, Serbia and Laos.

12  **Confusing, from an avalanche...** Authoritative (science based) writing conveying the 'children-grandchildren' emphasis is available (2009) from James Hansen's *Storms of My Grandchildren*. Most convincing from Hansen is his 1988 prediction that sometime in the first decade of the Millennium, Washington City would experience nine days each year with temperatures in excess of 35°C (a rarity in the 1980s). In fact in 2012 there were 23 days with temperatures in excess of 35°C.

13  **Climatic seasons set...** In temperate lands the traditional four seasons; in the humid tropics pretty much two; for Aboriginal and Maori as many as seven. Barber's 2007 title is *Swallow Citizens Whole*.

14  **War famine and disease...** The 1914–18 Great War was the last major conflict between warriors with warrior fatalities far exceeding civilian casualties. In all succeeding conflicts more that 50% of the fatalities have been collateral bystander and targeted civilian populations.

15  **If we lack...** One nation, Cuba, has already suffered from, and adjusted to, an 'artificial' Peak Oil event as a consequence of the 1970 US oil and trade embargo.

16  **We seem largely to know...** What Heller actually wrote in 1961, which I also pick up as relevant to these times, is that "There was only one catch and that was Catch-22, which specified that a concern for one's own safety in the face of dangers that were real and immediate was the product of a rational mind."

## CHAPTER 2: HUMANITY AND THE BIOSPHERE

17 **Three observations…** Neither Hawken (Cambridge astrophysicist and sometime mid-70s GradPad lunch-group member) or Dawkins (Oxford evolutionist) are connected with these assertions; but I fancy a speculative alignment. Hawken received an avalanche of little-hope-for-the-future presumption after seeking public opinion on the longer-term prospects for humankind. As for Dawkins, I doubt if he would reject the rationale of a evolutionary adjustment by a lessened population evolving 'backward' to survive.

18 **Naysayers unemcumbered…** Previous extinctions of species have occurred several times in the pre-civilisation past; most spectacularly as a result of the asteroid impact which, it is widely presumed, wiped out the dinosaurs. The current extinction event (2007 IUCN Red Listing) calculates 39% of species are threatened as a result of climate change, habitat destruction and degraded ecosystems; a pattern of extinction driven, this time, by human agency.

19 **Human dominance…** Nothing apocalyptic like a Doomsday Event (Harold Camping's Saturday May 11, 2011) is suggested; more a creepy degrading gradualism. Refer to Graig Dilworth's *Too Smart for Our Own Good*.

20 **Over recent decades…** From Japan's Ministry for Environment for the 2008 G8 (USA, UK, Germany, Italy, France, Japan, Canada, Russia). The greatest amounts of $CO_2$ emissions were then from: USA 23%, China 16%, Russia 6%, Japan 5%, India 4%.

21 **Poverty is a proportional…** Poverty is also a *relative* matter, for a television is a television whether it is a 10-inch black-and-white tube or a 100-inch plasma colour flatscreen; my reading being that in a general sense households with either are not poor. I assert that indigenous people, particularly those outside the money system, are correct in their lifestyle, and are in that sense 'rich' and sustainably worthy. The poor nations (the 'G77 plus China) comprises two-thirds of the world population; and overall they have done comparatively well by the environment in recent times.

   When it comes to handing over 'conscience money' from the rich to poor nation recipients, the record is shameful. The 2001 Bonn Declaration allocated $410 million a year for climate adaptation in developing countries. By 2009 this should have amounted to billions of dollars paid out, or in the pipeline. By late 2009 only $268 million had been allocated to developing nations.

22 **Only one-fifth of the global…** Further to my 'stone age poor' observation we have this from John Gray (2009) "… the project of promoting maximal economic growth is … the most vulgar ideal ever put before suffering humankind."

23 **Of the capitalist…** As an Assistant Director of Development Studies at Cambridge (1972–1984) my research and teaching responsibility was Land and Development within a trinity 'Economics, Sociology, Land Policy' with exposure to, and Third World empathy for, Marxian-styled analysis. My most thus-inclined writing at that time was *Ecodevelopment* (1981).

24 **Back to earth...** From Kunstler 2005 "Burning coal is still the greatest source of overall air pollution" — hence his adage 'keep coal in the hole'.

25 **Falloff in the production...** Peak Oil is the anticipated date at which global production 'peaks' prior to output declining everlastingly. Twenty-five books of concern were written on the topic in 2004–05; an authoritative and pointed example being Goodstein's *Out Of Gas*. In terms of climate change, more worrying than arrival of 'the Peak Oil day', is delayed arrival, with 'big oil' producers able to continue to meet demand from alternative carbon reserves, thereby further piling-up the anthropogenic greenhouse gas emissions. As of mid-2008 the BP 'Statistical Review of World Energy' contended that there was enough fossil carbon in reserve to maintain current levels of accelerating consumption well beyond 2050; in fact, capable of producing 2,500 gigatons of carbon dioxide, in a situation where burning an additional 565 gigatons would lift global temperatures above the two more degree 'tipping point'. This is severely disturbing. The worst case situation is where liquid oil demand is met from 'unconventional sources' (coal, tar sands, shale gas) producing aggregate emissions more substantial than those given off by petroleum oils.

26 **And in all this...** European Commission research obtained by EurActiv (2012) indicates that biodiesel from palm oil and soya approximates the pollution of fuel produced from tar sands fuel; and that ethanol from corn and maize, and sugar cane and beet, is about half as polluting as the orthodox production of fuel from crude oil. Biofuel produced from non-food plants grown on land unsuited to food-crop production, and biofuel obtained from commercial forestry discards, is more environmentally acceptable. But here a lingering problem remains, the amount of mechanical energy sequestered into production relative to the biofuel gain. Also there's Neil MacGregor's observation that "For some Mexicans it's unthinkable that maize, the divine food, should end up in a fuel tank."

27 **This style of living...** Here's a prescient bit of early cynicism from Lewis Mumford's *Myth of the Machine* (1934), about technological excess, particularly the automobile "There is only one efficient speed: faster; only one attractive destination: farther away; only one desirable size: bigger; only one rational quantitative goal: more."

28 **Suburbia is a long way...** In terms of mobility constraint: 1954: 50m cars; 1989: 350m cars; 1997: 500m cars. By 2010 globally there were 1bn vehicles (Society of Motor Manufacturers and Traders) well over half in the Americas and Europe. At any point in time, today, a million people are airborn.

29 **Adaptations can be made...** Rebecca Solnit *A Paradise Made in Hell: The Extraordinary Communities That Arise in Disaster*, 2009.

30 **Reason the situation...** Researchers at the Climate Dynamics Group (Oxford University) have calculated (2009) that half a trillion tonnes of fossil carbon has been consumed since the beginning of the industrial revolution — mostly in the last half century. This has been responsible for the recent +2° rise in average global temperature. The consumption of another half trillion tonnes will produce at least (in excess of!) another +2° rise in average global temperature.

## CHAPTER 3: CONSUMER CULTURE

31   **All this leads...** Most startling to me is Kate Raworth's finding (Worldwatch Institute *State of the World 2013* that "Just 11 percent of the global population generates about half of global $CO_2$ emissions."

32   **Carbon gas...** The 2009 reported level of global greenhouse gases was 385 parts per million (a 19% increase on 1970 levels) which is already higher than the ppm levels reached during the previous three interglacial epochs. The lag, then and now, between ppm peaking and full ice melt — culminating then in a four metre sea level rise — is not known; but it is a matter of centuries rather than decades. See the IPCC Working Group *Fifth Assessment Report 2013*. Another oddity is that as a result of the 2008 economic crisis (reduced fiscal discretionary spending in rich nations coupled to a reduced level of deforestation in poor nations) the Global Carbon Project (GCP) has assessed that global emissions of $CO_2$ actually fell 1.3% in 2009.

33   **Of course...** 390.9 ppm was reported in 2012 by the UNs World Meterological Organisation.

34   **Extrapolations establish...** What happened last time? Why did previous 'last time' climate changes occur without human agency? The answer lies with a flatter elliptical shape to the Earth's Orbit, guiding our planet closer to the sun, for several Millennia, about every 100,000 years. Also, intriguingly, overall temperature increase, even by a single degree globally, induces a massive volumetric expansion of the planet's water resource.

35   **Consideration moves...** 'Gaia' (Lovelock 2006) the "evolutionary [self regulating global] system in which any species, including humans, that persists with environmental changes that lessen the survival of its progeny, is doomed to extinction." Gaia is the metaphor employed by Lovelock, as a scientist, to profile the inter-relatedness of each to everything; to Lovelock the emergence of a Gaia cult is an unforeseen (and probably unwelcome) happening.

36   **No cosmic motive...** What Mark Lynas provides in *Six Degrees* is a researched summary perspective on the global outcome of climate change effects; by +1°C increments from One through to Six degrees — a harrowing progression. Each single degree step is presented incrementally, chapter by chapter, single degree by single degree.

37   **Ecological balance...** Mostly we select the United States as the prototypical consumerist La La Land; to my way of thinking the top spot goes to Switzerland.

38   **The rate of human...** This witticism is from Christopher Hitchens (2007): "My own view is that this planet is used as a penal colony, lunatic asylum and dumping ground by a superior civilisation, to get rid of the undesirable and unfit."

39   **All along...** The 'green-growth specialists' of the 1990s include Paul Hawken, Amory Lovins, Lester Brown, Jonathan Porrit and Andrew Szaz. Not all the 1970s writings were based on good science. Some of the polemic (notably Ehrlich on population) came to be considered over-the-top 'doomsdayish'.

40 **Shifts in the harnessing...** 'Biscuits for biscuits' — even worse, the global distribution of bottled water.

41 **Within all nations...** The OECD comprises around a fifth of the World Population, and generates half the global carbon dioxide emissions. In 2009 China eclipsed the USA as the leading $CO_2$ emitter.

42 **It is important...** The IQ quip is not a put down, for the IQ of faux wealth profiteers is undoubtedly well above average. In times past men and women with high IQs tended to flock to the Law, Medicine, Academe or Politics; from which platforms they secured a respectable and comfortable lifestyle. Now the mostly male high IQ achievers are involved in bonus-rewarding financial manipulation, from which they access all the lifestyle trappings of excess consumption. To my way of thinking there is a likely three-way correlation between high IQ maleness, perceptions of insecurity, and greed amounting to a sociopathic indifference to the consequences of climate change.

43 **The growth-on-growth...** The IPCC reports prior to the 2007 *Fourth Report* had to accommodate politically entrenched sub-texts. By 2007 they were aided by the computer power of data assembly and the logic of computer analysis. The account by Spencer Weart *The Discovery of Global Warming* provides insight into the 'behind the scenes' actions of IPCC supporters and detractors. The Berkeley Earth Surface Temperature project (BEST) 2012 exhibits an objectivity that has changed the minds of some previous naysayers.

44 **This reasoning...** In the early 1990s Francis Fukuyama provoked a great deal of debate with his *End of History* 1992. John Gray's 2013 tile *The Silence of Animals: On Progress and Other Modern Myths* erodes human hope in a shift toward Fukuyuma's fatalism. It's not an attitude I am at ease with.

## CHAPTER 4: ENVIRONMENTAL AWARENESS

45 **Knowing where matters...** Fascinating and informative historical data on the environment of the past is coming through from fossil ice core analyses. View the 2013 documentary *Thin Ice* and go to www.thiniceclimate.org.

46 **The popular media...** In a chapter 'The Democratic Deficit' Tony Judt (*Ill Fares the Land* 2010) questions the worth of electronic contact "... even if the students of Berkeley, Berlin and Bangalore share a common set of interests, these do not translate into community."

47 **Flawed reasoning...** There can arise an inefficiency relative to local production and sales. As an example: consumers motoring across a city to a Farmer's Market contrasts with the efficiency of a truck taking market garden produce in bulk to a local supermarket.

48 **The media has an easy...** The high priests of the appealing 'footprint' explanation, as previously noted, are Wackernagel and Rees and their seminal text (1996) *Our Ecological Footprint*. The site www.waterfootprint.org claims that the production of 1kg of beef milk and wheat requires, respectively, 16,000, 1000 and 1350 litres of water.

49 **More reprehensible...** A paradox. Some among those of outstanding wealth

and privilege — sports stars, celebrity actors and a few super-rich — are also outstanding ambassadors for biospheric good health.

50 **Eco-labelling...** Sequestration? This term, in common eco-environmental usage does not appear in the eco-environmental dictionaries consulted. There are legal and chemistry meanings to the word, purloined by environmentalists. In eco-environmental usage sequestered energy is the captured and maybe stored energy, particularly from fossil carbon stock, put into (embedded in) the manufacture of a consumer durable, or fertiliser, or holiday (as examples) at, it is claimed by the Ayres brothers (2010) a rate of 13% gain and 87% not used. The expression 'to sequestrate' is in use to describe permanently storing away nuclear, toxic and excess $CO_2$ waste.

51 **A logical extension...** Some design parameters for an environmentally considerate prison are available from *Prison Architecture* by Leslie Fairweather and Sean McConville, Architectural Press, 2000.

52 **Advertising fuels...** It is important to avoid romanticising 'localism'. The process of 'living lightly locally' involves sacrifice, with some loss of consumer comfort and much dedication to principle. Kropotkin was published in 1889, and is available in a 1985 reprint *Fields Factories and Workshops Tomorrow*. In her *Paradise Built in Hell* Rebecca Solnit (2009) establishes a contemporary association with these ideals.

53 **Additional hiearchical...** David Owen former leader of the Social Democratic Party in Britain, and by profession a psychiatrist, identifies and admits to the 'political hubris syndrome' as an occupational hazard for senior parliamentarians.

54 **There are two other...** The New Economics Foundation has proposed that jurisdictions of wealth should move slowly toward a 21 hour working week, presumably over three or four days.

55 **A complication arises...** There are 'capital adequacy rules' under an international convention (Basel I and Basel II 2007) which set out to establish the amount of lending which institutions, mainly banks, have to hold as 'cover' in relation to the loans they make. This is a recipe for disaster when an economy falters for any reason and there's a run on deposits. In September 2010 Basel III proposed the slow introduction of a larger amount of top-quality Tier-1 capital for banks to hold in the future — a mere 7% in total! The well understood Basel Rules have always been assiduously observed, in the breach!

56 **The recurring problems...** Profiting from the 'tokenism of biofuel production' profiled by a 2008 'splash and dash' scam. European product was shipped to the United States where a 'dash' of local biofuel secured a hefty Stateside subsidy for the whole load, which was then shipped back to Europe for on-sale as biofuel — at vast overall emissions 'cost' to the environment, and considerable fiscal profit to the entrepreneurs. Source: European Biodiesel Board. (EBB 2008).

57 **Taxing carbon consumption...** The purest option would be electric powered vehicles recharged with renewable (wind water and solar) energy. The 2007 US

Federal Energy Bill first set out to raise the Corporate Average Fuel Economy to 15 km/L by 2020 as a first step improvement to national vehicle mileage efficiency.

58 **Laudable but less assuring...** The 'get with the game', 'do your bit', 'enlist with the team' gradualism is becoming known, pitiably, as the 'Sokolow wedge' — consult Sokolow in the References.

## CHAPTER 5: PEOPLE AND THE ENVIRONMENT

59 **During the low-tech...** Telling, personally, is the fact that global population at two billion around the time of my early childhood had doubled to four billion before I gained the right to vote, and is set to double again to eight billion as I prepare to depart this mortal coil: 2 - 4 - 8 - ?

60 **What is proving...** Of concern is the recent expansion of wealth and consumption in Asia (surging carbon uptake and luxury purchases) leading to an overburdening waste. An economic growth model disdained by Chandran Nair in *Consumptionomics* (2011).

61 **We now know...** The phrasing 'The limits to growth' is the title given by the Meadows' husband and wife team, with others, to their 1972 book; revived in 2004 with 'a thirty years update'.

62 **None of these reductions...** The surge of wealth in China is inducing a concern about future labour supply and a review of the one-child policy. In contrast, poverty in highland Uganda and Tanzania is inducing people to again have larger families for their geriatric support.

63 **This leads to...** In terms of carbon emissions Stern (2009) puts Settler Societies at >20 tonnes per capita annually and sub-Saharan Africa at around one tonne per capita. Only 3% of the planetary water resource is non-saline; and over two-thirds of that 3% is currently icebound. The vast oceanic and atmospheric 'sinks' balance the biospheric living web.

64 **As noted earlier...** It is popularly held that rural poverty is more wholesome and easily endured than urban poverty. Urban residential creates societies with an urban structure pretty much compartmentalised away from its food fibre and materials supply system.

## CHAPTER 6: TECHNOLOGY AND THE ENVIRONMENT

65 **In my small community...** 'Environmental abuse' is an 'externality' (an unattended-to and unpaid-for adverse effect) which results in the community at large suffering the consequence of individual, institutional and corporate discard, damage or neglect.

66 **Why me or us?...** New Zealand has been my home and academic base over the last 25 years. The less than 0.1% fraction of overall population is responsible for 0.2% of global emissions — a per capita rate double the global average. Some suggest that the minor global proportions of population and emissions, and the nation's isolation, exempt New Zealand from an obligation to reduce its pattern of excess emissions. But sea level rise, average temperature rise and

increased atmospheric $CO_2$ persist as a 100% local as well as a 100% global calamity.

67 **Contrarian propaganda...** The US-centred Heartland Institute is a group of contrarians dedicated to discrediting climate change science. They have been liberally funded by wealthy private sceptics, and tobacco petroleum and liquor companies, with some recent falloff in support from major corporates.

68 **Inattentional blindness...** A prominent sceptic is Ian Plimer (2009) author of *Heaven and Earth*. An account of 'inattentional blindness' is *The Invisible Gorilla* by Daniel Simons and Christopher Chabris (2010).

69 **Out of such inefficiency...** Efficient energy capture and energy use is a core economic advocacy put out by Robert Ayres and his brother Edward who produced the '13 percent' figure. (*Crossing the Energy Divide*, 2009)

70 **The forthcoming...** My 2004 *Sustainable Urban Planning* Box 3.4 — Soft Pathways has been sourced.

71 **A society rendered dysfunctional...** China and India lead all other nations with their volumes of coal consumption, with the World's largest untapped coal deposit awaiting exploitation in nearby Mongolia.

72 **What is the most optimistic...** Carbon capture and storage (CCS) involves the physical 'permanent' storage of captured carbon dioxide deep underground or deep underwater, and is reliant on producing and using even more energy to do this to operate the technology. Some of the best research work is being done in Victoria (Australia) while the supplier of brown coal, NSW and Queensland (also Australia) blithely exports to the now greatest $CO_2$ producer, China! Refer also to work done at the Hadley Centre For Climate Prediction; Richard Betts/Peter Cox, 2007.

73 **This leads to...** In Brazil, the environmental injustice of tract felled Amazonian and North East forests, once essential carbon sinks for $CO_2$ absorption, is exacerbated by using the newly deforested landscapes for soft-crop ethanol production. An Australian company, Licella, is developing (2012) a commercially viable catalytic hydro-thermal reactor.

74 **More harebrained...** A positive take on TPD, involving the use of putrescible sewage waste and the possibility of $CO_2$ recycling, is explored in Chapter 14 (Targets).

75 **There are also nuclear options...** The Pebble Bed system is designed to be safer and smaller (modular). However, it is a system still at the prototype stage, and suffers from escalating development costs. And, also, for all manner of nuclear plants, there is a need to factor in the considerable carbon cost of mining and delivering uranium, and the carbon costs involved in dismantling plants after their use-by date.

76 **Some jurisdictions reject...** The nuclear option is controversial. Paul Steele, Dalesman, former student, valued confidant and lifelong friend, wrote "I have one big disagreement with you, and that concerns nuclear power. My guess is that you like it as little as I do, but your writing does endorse it, maybe not with enthusiasm. It's really just another part of the materialist culture that's brought us to where we are."

77  **The use of plastics...** Pre-Olympics 2008 hype? China, the World leader in plastic bag production, banned their use. More to the point would be doing away with 80% of the plastic and paper wrapping which envelopes every manufactured item to come out of China, as though the manufacturer's of plastic baubles were dispatching some of the nation's Ming Vases!

78  **Food production...** Bulk transportation in large container shipping vessels, now (2013) almost 400 metres long, and by lengthy convoy trains is, relative to truck-container transportation, energy efficient.

79  **The technologies engaged...** In the past our ancestor's food supply inheritance comprised the soils and aquifers now widely exploited to a state of reduced utility — progressively exhausted, eroded, desiccated, salinated and toxin laden.

80  **Adjacent to feral fish...** I'm not making a case against total meat intake. But I would argue for dietary meat displacement partly on personal health grounds, and because the equivalent of eight kilos of pastoral and grain feed (and bathtubs of water!) goes into the usually highly fossil fuelled production of a kilo of animal protein food. So, therefore, small meat portions. Also: farmed fish before fowl (chicken), and pork and mutton before beef; and sail-assisted feral fishing. In wealthy societies vast quantities of still edible 'fresh food' gets binned in supermarkets prior to the 'sell by date'.

81  **Most of the non-solar...** Ian Henderson, Unitec colleague, comments: "My sister in the West Country (England) produces water from her own well. Everybody wants to buy local, so her neighbours buy her water after it has been to a storage and distribution centre in the Midlands [200 miles away!]" — food calorie value zero, industrial calorie input unconscionable!

82  **Citizens...** Joseph Adelson, *Inventing Adolescence* 1986 alludes to the "idyll of suburban domesticity" which I code as: plot house cars gizmos supermarketing jobs children.

83  **The Actions and Adaptations...** Calculation of carbon footprinting is available at www.climatecare.org with leads to other options.

## CHAPTER 7: BIOSPHERIC ECONOMICS

84  **Acting against the collective...** The uptake of mineral carbons is not dominantly a matter of availability and cost at the gas pump. The western suburbanite family consumes more mineral carbon for the delivery of food onto their plate, and for home heating and maintanance, than is consumed by their use of a car. Consult Eric Zency 'Energy as Master Resource' in the Worldwatch Institute (2013) *Is Sustainability still possible?*

85  **From the dawn...** For an account of early economic behaviour consult Marshall Sahlin's *Stone Age Economics* 1972.

86  **Clearly there is...** An irony: banks created to avert financial panic came to induce unprecedented financial chaos.

87  **One hundred years ago...** WWI, followed notably by WWII, destroyed much of the physical infrastructure of the nations of wealth. Sixty years of relative

global peace since WWII and lesser skirmishing in poorer nations, has abetted the assembly of mega-wealth and a rapid accretion of population.

88  **Much more than was…** The failure of democracy for the victimised poor is addressed in Paul Collier's *The Bottom Billion*, 2006; and in the New Internationalist 'No-nonsense Guide' *Democracy* by Richard Swift (2009 edition).

89  **All along…** Also consider, along with the reasoning in this paragraph, the 'selfish gene' rationale driving financiers to risk pyramid gaming where, when they 'won' they could walk away with their 'winnings' personally well clear of the mayhem left in their wake when they 'lost'.

90  **The worst of it…** For the United States, a national debt equivalent to $30,000 of liability, and growing, for each individual person. Source: 2008 film by Patrick Creadon I.O.U.S.A.

91  **The early 20th century…** We have this from Bruce Jesson (1999) "Left to itself finance cannot develop a culture that goes beyond greed and egocentric individualism."

92  **Paradigmatic resource looting…** Capital adequacy rules [the Basel I (1968) and II (2004) guidelines] were supposed to be about the leverage of bank and lending institution holdings relative to the loans they made. It varied, but 8% of capital held in reserve was a consistent figure (around 1:12) which would seem pretty outrageous to ordinary mortals. To investment banks this constraining ratio was merely a challenge to get around. Prior to the 2008 meltdown some lenders in the US managed to issue loans in excess of their holdings by as much as 35 to 1 — all the while squirrelling away a significant proportion of their 'successful' risk-taking business as enormous bonus 'entitlements'.

93  **The leverage which…** The zenith of socially obscene and environmentally damaging phoney liquidity creation is-was the Lehman Banking system of loan generation, claimed by some observers to be in excess of fifty times the extent of the bank's capital holding. Well prior to the Lehman 'social obscenity' J K Galbraith observed that "The process by which banks create money is so simple that the mind is repelled."

94  **The linked together…** The 'Washington Consensus' under Ronald Reagan with support from Margaret Thatcher discouraged government intervention. It held out for trade liberalisation, privatisation, corporatisation and market deregulation. The US Government is now (from 2011) served by a Financial Stability Oversight Council staffed in the main by Reagan-Clinton era advisors! For a more sophisticated take on the 'Fiscal Alchemy' caricatured in the aside read Chapter 4 (Enter the Geniuses) in Johns Lanchester's *Whoops* which tracks the work of David Li on the copula function and risk diversion. The authority on this and the Black-Scholes (plus Merton) equation governing the ephemeral trading of investments in investments is Ian Stewart's *17 Equations That Changed the World*. In a context where very few understand what's going on, Stewart argues for a radical overhaul involving 'more mathematics, not less'!

95  **We look for culprits…** According to Greenspan (2009) 'A staggering equity loss of well over $40 trillion'. The film 'Inside Job' is revealing; the text *Tangled Webs* by James Stewart (2011) portrays four examples of gross deceit.

96  **Furthermore…** A colourful 'factoral' account of the the the frat-dorm 'gaming' world of the Washington Consensus and the Wall Street co-community around the time of the 2007–08 fiscal implosion is given in Sorkin's *Too Big To Fail*. And there's a movie, Charles Ferguson's lucid 'Inside Job' (2011).

97  **My personal turning…** The Kashmir project formed a chapter 'How do you justify conservation in a developing country?' in *Conservation of the Indian Heritage*, Cosmo Press, 1989.

98  **Within democratic…** Consider Africa; the vegetable and protein supplier, mineral provider, timber provisioner, labour source, toxic waste dump, even flower vase filler of first resort. Initially colonised in the main by the Anglo-French, now economically colonised by the Euro Zone nations and China.

99  **In the breech…** This, from Steven Pearlstein of the *Washington Post* (November 7th 2011) — "Silly you. You actually thought companies existed to make products and profits… .You weren't sophisticated enough to realize that these are just different 'asset classes' meant to give investors something to speculate in … [as] futures, swaps and other derivative instruments."

100 **At a local…** For all the finger pointing over the 2010 Gulf of Mexico oil spill, the rate of discharge and extent of socio-environmental damage there, is less than the oil-spill desecration of the Niger Delta. The perpetrators of that disaster, at home and abroad, would about top my list to prosecute for ecocide in an International Criminal Court for Environmental Justice; with a potential to be eclipsed by the foreseeable climate changing potential arising out of heavy pollution from exploiting the vast Canadian tar-sands.

The Crimes Against Peace include: Genocide, Against Humanity, War Crimes, Aggression; with the now suggested addition of Ecocide.

## CHAPTER 8: PART A OVERVIEW

101 **An expansion of…** As reflective white ice gives way to heat absorbing dark water and rock, an 'albedo flip' phenomenon is triggered, leading to slow yet irreversible ice-field meltdown. There could be, to one way of reasoning, an analogy here with a probable 'economic albedo flip' when confidence in the worth of money as a store of value dissipates, globally, without recovery.

102 **Perversely…** Here are three sharp commentaries on the pitfalls of modern living: *Affluenza* by Oliver James, *Bait and Switch* by Barbara Ehrenreich (also her earlier *Fear of Falling*), and *The Spirit Level* by Wilkinson and Pickett.

103 **For the majority of us…** 'Man(!) Thinks' — Benedictus de Spinoza was of course writing in the 17th century.

## CHAPTER 9: NEW PARADIGM

104 **What may be a surprise…** The wealth-against-poor significance of the Western way of thinking about who are the 'good guys' and who are the 'bad

guys' in terms of environmental sustainment, comes through from a joint Yale-Columbia study. Some 140 nations were ranked; and I have accessed a 2008 listing of the top-ranked third. That top 'best-behaved most-sustainable fifty' included only three poor nations: Costa Rica (5th), Ecuador (22nd) and Dominican Republic (33rd). I would wager that there are fifty alternative poor nations where the levels of sustainability per capita is in fact better than the top ranked most sustainable fifty listed in the Yale-Columbia study!

105 **To this confusion...** Within democracies of wealth, systemic failure arises from 'legitimate' lobbying — protecting interests and avoiding disclosure.

106 **Hyper growth...** A few months after first writing this paragraph I was invited by an academic visitor to supper in a revolving restaurant atop a casino tower in their hotel; an unusually contrived way to eat fish! Later, strolling through the Gaming Hall, I got to thinking of Foucault and his interpretation of societies through the prism of asylums and penitentiaries. The casino experience could be interpreted as a way to understand the contrived values of modern society in terms of 'opportunistic work' in money-manipulating employment, and in 'opportunistic play' at the casino.

In another situation of make-believe-meeting-reality, the young actor Shia LaBeouf boning-up for the 2010 sequel to Oliver Stone's 1987 'Wall Street', worked on the trading floor at Goldman Sachs and other hedge-fund people, opening an account with $20,000. In short order this parlayed into $297,000. Instinctively most of us would feel envy; collectively we should register disgust that this socially unfair travesty could legitimately take place.

107 **Some Creationist's claim...** How come Adam and Eve are always shown to have navel buttons?

108 **Beyond the political...** The buy-out approach to conservancy has been assessed as a modest 2.4% of current global GNP. The '2.4% of GNP' comes from Jeffrey Sachs *Common Wealth* (2008). A whacking great 'trillion dollars a year' through to 2050 is a figure suggested by the International Energy Agency (2008) and relates to a quartering of emissions from all mineral carbon users. A query arises from both estimates: would a per capita quartering of emissions (mainly through fuel efficiency, decarbonising processes, benign electricity provision) be enough, for this reckoning doesn't factor-in the likely extra billions of consumer population.

109 **Pan-political agreement...** Anthony Giddens *Politics of Climate Change*, 2009.

110 **In that curious...** The "average citizen of the United States, Europe and Japan consumes 32 times more resources ... and puts out 32 times more waste, than do inhabitants of the Third World." (Diamond 2005).

111 **Adaptation is the task...** For the de-growth advocacy consult Serge Latouche, *La Decroissance* the magazine: also at www.ladecroissance.net.

112 **Hovering around the...** Elinor Ostrom, shared 2009 Nobel Prize in Economics. Her project supported the principle of 'subsidiarity' for harnessing the creativity, and exercising the authority of local communities, to sustainably manage resource use.

113 **Looking-in as it were...** To Wittgenstein the bounty of life was simply an inexplicable 'wonderment'. Annie March corresponds from Tasmania "What mind boggling sophistication keeps trees aloft, water sweetly flowing, and gets oaks into acorns and back out again?" For Dworkin, human living amounts to a 'successful performance' bound for eventual extinction.

## CHAPTER 10: INFORMATION & EDUCATION

114 **The flow of information...** Ian Henderson again "This malaise of realisation is contributed to by us all. We still buy into climbing the economic 'security' ladder. Even those of us who accept the need for action (a small minority) have no clue about the use of resources, our carbon footprint, or what it might take to do anything about it".

115 **Strategists...** Latching on to the slightest IPCC pratfall is a naysayer favourite. Contrary to knowledge that in some temperate zone contexts glaciers were actually pushing their terminal face further downhill as a consequence of increased high-altitude snowfalls, one of the 2007 IPCC reports incorporated an unfounded, generalised and converse contra-opinion about universal glacier retreat from non-verified sources, to the delight of the climate change skeptics.

116 **Hard facts...** The $CO_2$ absorbtive capacity of tropical rainforests (Amazon Congo Burma-to-Borneo) is roughly five times the rate of temperate-zone rainforests and ten times that of high latitude forests.

117 **As noted previously...** Stephen Schneider writes ('Confidence Consensus and Uncertainty' 2010) with authority about the degree of objectivity underlying the IPCC consensus.

118 **Secret knowledge...** The most significant precursor to profits taken from tobacco, despite the known ill-health effects, being the cynical (and very profitable) addition of lead to petrol.

119 **A GlobalGood...** It is relevant to note that the IPCC, an authoritative body working out of dedicated facilities, does not carry on its own research, nor does it specifically monitor actual climate change or related phenomenon. There is a case for the IPCC to be charged with a future remit to benchmark and disseminate the levels of sustainability (in a sense 'footprinting') appropriate to working, schooling, housing, feeding and mobility patterns.

120 **There is a reluctance...** The early (1989) multi-authored and influential Worldwatch Institute Report(s) on the *State of the World* conveyed nothing substantial on either mass wealth or mass population as having an adverse influence on climate change: both themes central to this writing.

121 **The point being established...** 'Transparency, honest dealing, disclosure' would be hugely assisted by the adoption in all nations of high quality Extensible Business Reporting Language (XBRL) of the kind promoted by the International Accounting Standards Committee.

122 **Fired by turbo-wealth...** Mineral carbon (notably liquid oil) depletion, and the related Peak Oil phenomenon leading to increased dirty carbon

uptake, are to the fore in terms of ambient carbon gas content. But almost as significant, certainly very important for the future support and wellbeing of humankind, is soil depletion, soil erosion, soil salination, aquifer depletion, seawater quality and species extinction.

123 **Supercomputors exist...** Beyond supercomputers several really superduper stand-alone facilities have been assembled: in the US (Blue Gene, Roadrunner then in 2010 Cray XT5), UK (Hector),and the largest (Tianhe-1A 2010) in China. Their first major use is usually military. These behemoths are very attractive to those who seek an advantage in identifying profitable investment strategies, in a flash, before their competitors. Each massive installation has only three, maybe five, years of operational effectiveness.

124 **GlobalGood...** A further example of potentially effective emissions tracking is the Vulcan system being developed at Purdue University, Indiana. This technology has the potential to pinpoint local $CO_2$ emissions over a whole country hour-by-hour. The capability of such a system to first detect, then the possibility of enforcement and cleanup is promising in terms of the GlobalGood recommendation.

## CHAPTER 11: ECONOMICS & DEMOGRAPHICS

125 **Exhortations to turn off...** In his *Sustainable Energy Without the Hot Air* David Mackay "...put it this way: the energy saved in switching off your phone charger for one day is used up in one second of car driving."

126 **Wealth accumulation...** A financial appeasement arrangement, for wealthy nations to aid the rectification of habitat damage in poor nations, was cobbled together in 2001 as the Bonn Declaration. The deal was for a consortium of wealthy nations to provide $410m each year to poorer nations for 'climatic adaptations'; compensating for the rich nation's damage to the poor nation's habitat. By the end of the decade only $268m out of the billion-plus dollars pledged by the holders of wealth can be accounted for. Over the same period the cost of the largely futile 'coalition' intervention in Iraq exceeded one trillion dollars!

127 **Population number reduction...** A likely first abstention in a time of hyperinflation-recession and massive unemployment will be air travel; then out-of-season food supplies, automobile overuse, and top-end medical services.

128 **For most of us...** The key factors affecting future mineral carbon supply and carbon emissions are: 1) Increased demand from 'emerging' poorer economies 2) A dwindling of the discovery and supply of liquid carbons 3) Continued operation of a 'free' demand-supply market in 'dirty' carbon products 4) hydraulic fracturing to release shale gas. The de-wealthing (stabilisation) focus would mainly be on the reduction of volatile liquidity — poleaxing the derivatives trade, banning scams for securitising debt, and ensuring that government treasuries curtail the 'printing' of money. See, also, the 'Economic Stability and Resilience' sidebar coming-up later in this chapter.

129 **Nations of wealth...** For more data and refined reasoning consult David Satterthwait's 'Implications of Population Growth...' (2009).

130 **The poorer four...** There is a literary analogy from the Portuguese novelist Jose Saramango (1995); his harrowing tale of a community struggling with affliction and adjustment in a situation of severe adversity: about which, at his Nobel prize giving, he observed "... man stopped respecting himself when he lost the respect due to [all other] fellow creatures."

131 **Here arises...** Support and argument for a steady state fiscal system (also known as decroissance) is made by the Centre for the Advancement of the Steady State Economy (CASSE — www.steadystate.org.)

132 **Monetary Services...** The most prominent gain-playing is 'flash trading' by individuals in collusion with prominent brokerage firms, using mega-computor installations to execute superfast buy-and-sell trades (over 60% of US stock market activity in 2010) in a 'game' of no community worth which indulges armchair 'players'. This contributes hugely, as a consumer income accelerant, to global resource plunder, thence carbon effusions and toxic overload.

From *Whoops!* (2010) here is John Lanchester: "Imagine if you go to a casino and play roulette, and every time you won you kept the money... you keep all the winnings. Whenever the wheel lands on zero, you lose ... but the good news is, you don't have to pay: the government steps in and picks up the tab, and you keep your previous winnings."

133 **The lesson learnt...** October 2009: two previous Bear Sterns hedge fund managers, accused of luring investors into a market they knew was collapsing, faced a jury trial. Obfuscation resulted in pretty-much a let-off!

134 **The laissez faire expansion...** Parecon has four characteristics. 'Solidarity' can be calibrated in this text with stability; 'self-management' aligns with subsidiarity; 'diversity' can be envisaged as scaling back to living-lightly-locally; 'equity' with evenness and balance within communities.

135 **The Macro, Correcting and Control...** Relative to this call for New Economic Stability was the 1975 call from the UN for a *New International Economic Order*. A compassionate and moral underpinning for both economic 'stability' and 'order', is given by Amartya Sen *The Idea of Justice*.

136 **In contrast...** I've been over the core of this reasoning before at greater length — my chapter on 'Practice Ethics' in *Sustainable Urban Planning*, 2004.

137 **In defense...** Also, how about retro-pledging?

138 **Also telling...** My first stand-alone writing (*Ecodevelopment* 1981) listed eleven 'Macro Principles' which reach toward a 'balanced' convergence:— Ideological Commitment, Political and Administrative Integrity, International Parity, Alleviation of Poverty-Hunger, Eradication of Disease-Misery, Arms Reduction, Self Sufficiency, Reduction of Urban Squalor, Balanced Human Populations, Resource Conservation, Environmental Protection.

139 **If, as projected by WHO...** United Nations projections put the 2050 world population at 9.2 billion. The reason why population might well grow to this enormity, centres in part on the perceived need for the poor and uneducated

populations to provide support persons for those in their old age.

140 **Another community...** The prime co-housing reference is McCamant and Durrett 1994.

## CHAPTER 12: RESOURCE CONSERVANCY WITH WASTE CONSERVANCY

141 **Another way...** The Planetary Boundary people (Nature Vol 461 Sept 2009) argue that the fixing of atmospheric nitrogen, which ends up as nitrates in waterways and the ocean, is as equally serious and challenging as carbon gas effusions.

A cautionary consideration arises in that the manufacture and installation of solar panels and wind turbines involves the considerable embodiment of mineral carbon energy; and although this is gainfully assuaged during the product's lifetime of use, it must of course be factored-in.

142 **Think now in terms...** The 'contraction and convergence' mantra (managing efficiencies 'up' and inefficiencies 'down') focuses on the regulation of personal carbon dioxide output (which involves also 'crediting' effective carbon dioxide sequestration). It is a concept which works on the basis that each person (consumer unit) is allotted a non-tradeable (personally disposable) quota of carbon dioxide emission; and the surcharging-of taxing-on and penalising-of excess carbon consumption. Problem is, can this theoretical approach be fashioned into something enforceable?

143 **Staying with contraction...** Globally there is one domesticated ruminant animal to every two persons, with each animal producing methane 30 times more atmospherically degrading than carbon dioxide, and aggregating as 14.5% of humanly 'produced' (controlled) greenhouse gas emissions. Phasing-out domesticated ruminants by say 50% would be hugely beneficial to atmospheric wellbeing.

144 **In times of fiscal buoyancy...** What some governments contemplate doing with environmental levies and penalty income is to grant these monies to those of poverty; increasing their mobility and consumption, and their production of carbon gases, again with no net gain to the stability of the biosphere.

145 **In order to work...** The noble sounding International Emissions Trading Association (IETA) exists as a vehicle for the likes of oil companies, mining conglomerates and cement manufacturers to negotiate carbon gas offsetting, and to operate their, to them, all-important carbon trading for profit.

146 **A first problem...** At boom time early 2008 a tonne of carbon emissions traded on the carbon market for $40. By deep-recession a year later a tonne could be bought for $10!

147 **The greater need...** During mid-1974 I spent twelve weeks assessing development impacts in the Pre-Amazon, Rio Negro and Amazon river systems. In Pre-Amazonia it had been already established that the soils were thin and the rainfall marginal for forest support, and that desertification

toward a savannah landscape would follow forest clearance within a decade. Inland from Santarem to Ruralopolis and around Tefe and Iquitos, all on the Amazon, the landscape for over a mile either side of the new highways were already cleared of forest. At that time the Rio Negro was unroaded and the river ran dark and clear. Further to the overall theme of this paragraph: there are half a dozen carbon dioxide absorber ideas; synthetic trees, mirrors in space, dust blanketing, cloud seeding, permanent carbon fixing etc — Mark Jaccard (2006) *Sustainable Fossil Fuels*.

148 **Localisation of government...** Sourcing locally (in a less mechanised way) is even more significant (carbon saving) than simply 'buying green'.

149 **We can add...** In his *Maybe One* Bill McKibben quotes Cornell's David Pimentel "A head of lettuce is 95% water, just 50 calories of energy, but it takes 400 calories of energy to grow it in California ... and another 1,800 to ship it East."

150 **Another item to consider...** Local farm gate and front door sales of food fibre and construction materials need not, in general, be burdened with GreenTick information.

151 **Input-output spreadsheet modellers...** In the input-output context of money, the ratio of asset value to lending amount can be set, then reduced or increased during times of growth or recession, with the all-round consequences reflected throughout the rest of the Input-Output table. In this way adverse environmental changes of the kind that occurred as a direct result of the accelerated increase of liquidity during the last thirty years, can be foretold. *My Sustainable Urban Planning* (Ch 4 Growth Pattern Management: Box 4 Input-Output Patterns) profiles a schematic input-output situation.

152 **Economic, population, consumption...** Relative to the matter of 'stabilis(ing) or reduc(ing) the loss of biodiversity' delegates at the United Nation's 2010 biodiversity conference adopted a protocol (the Aichi Targets) to halve the loss of natural habitats and expand nature reserves to cover 17% of the world's land area; with effect from 2020 (!). These are sentiments to applaud; but in terms of realism, what planet are these 'delegates' advising?

153 **Input-output displays...** Economic recessions consistently track oil prices. On future occasions when this happens a recession could be dampened — at least over the short term — by investing much of the petro-currency accumulation, Keynesian style, in environmental good-cause projects. The challenge, with the approach of a moribund economic slump, being to plan for and to manage the let-down in a manner which is socially and environmentally least hurtful.

154 **It is a matter of...** Regarding 'sense of place'. My first academic appointment was to a lectureship at the School of Town and Country Planning within the University of Newcastle upon Tyne. Part of my brief was to restart the house journal *Planning Outlook*. The journal took its title from Patrick Geddes' 'Outlook Tower' in Old Town Edinburgh; a camera obscura installed by Geddes early in the 20th century to encourage citizens to embrace his belief in the significance of 'place work folk' in community life.

## CHAPTER 13: GOALS

155 **Eco-advice emanates...** The matter of ethical (responsible) investment along with the poleaxing of 'tax havening' illustrates a regard for conserving (stabilising) wealth along with a conservation of the environment — doing the right thing. The UN has established a set of 'responsible investment principles' that encompass environmental social and ethical concerns (ESGs). The majority of ESG participants are reportedly doing well, as investors!

156 **Solutions to climate change...** An extensive writing, but daunting, is the 700 page *Environmental Science* (McKinney and Schoch, 1998). An extreme example of wealthy consumer gullibility is the biodegradable golf tee. Golf (my sometime recreational pastime) is a profligate consumer of two of the World's most important resources, water and mineral carbons; and on this account has been criticised as a grotesque environmental travesty. I am now able to ameliorate this travesty by sourcing biodegradable tees! Shall I drive over to Silverdale to purchase a packet; or would it be more environmentally friendly to have them sent by Courier?

157 **The one-fifth of World population...** The 'Settler Societies', harking back in origin to the British Isles are Canada, Australia, New Zealand and the United States. Of course the four-fifths relatively poor and relatively non-polluting global population equally (although disproportionately) share the the global-wide pollution burden.

158 **My working life...** The 'deliberate fully consulted village plan' (see also Oceanic Exemplar II — the Natongandravu account) was an attempt to fully integrate with, and plan for, a discreet hamlet in every physical, social and economic detail. It was exacting, sometimes exhausting, often exhilarating; but also hugely time consuming. It was found impossible to extrapolate this experience into orthodox planning methodology!

159 **Planning processes...** 'Strategic withdrawal' and 'conflict avoidance' are well established military procedures, now part of corporate governance.

160 **Envisage caring sharing...** Electronic systems can be fingered for getting us into much of the current climatic pickle and biospheric breakdown; yet, an irony, they could well prove to be the information help-line and decision-taking mode for getting us out of that pickle and breakdown.

161 **The other significant...** India has had a renewable energy programme since Stockholm (1972) and is a significant contributor to the manufacture of technical components for environmentally sound consumer durables and the likes of wind turbines. China, as noted, is making good progress; driven largely by the simple fact that in terms of both sea level rise, and the advance of desertification, it is vulnerable and has the most to lose.

162 **Returning to the tracking...** The proposed *Kyoto2* protocol is outlined in a book of that title by Oliver Tickell (2008). It is designed to be driven by market mechanisms. The potential is impressive; the challenge gargantuan: the opportunities for the process to be commercially commodified considerable.

163 **An accessory to consumption...** Japan's Trade Ministry, conscience pricked

by the Kyoto Initiative, announced (September 2008) a scheme for displaying the carbon dioxide accumulated by a product during manufacture distribution and disposal. Initially the scheme is to be voluntary; for packaged food, detergents and electrical goods.

164 **Food supply...** Germany, Spain, Portugal and Denmark (spectacularly offshore Denmark) have achieved the greatest proportion (relative to population) of artificial solar capture and wind energy contribution to national grids in Europe (advocated by Herman Sheer, German MP; and now known widely as Scheer's Law). Other 'solar' systems include hydro-electric generation, with considerable potential available from 'run of river' installations. Tapping the huge potential of tidal and wave systems is mostly still at the design stage. Solar cell and heat cell collectors are suited to in situ circumstances, and are still undergoing design, development and cost reduction. The output efficacy of solar-supported and natural systems, relative to their cost of installation and length-of-life, ranks top-down: hydro, wind, tidal and wave, battery and heat cell capture, biofuels. There is some potential for geothermal heat exchange.

165 **An expanding population...** With regard to the effectiveness of fertility management (primarily birth limitation) the Optimum Population Trust at the London School of Economics calculated (2009) that in monetary terms every dollar spent on family planning is almost equivalent to the environmental benefit of five dollars spent on conventional green technologies.

166 **The MUD concept...** Wackernagle and Rees (1996) reason that higher urban densities tend to shrink the Ecological Footprint. The optimal eco-village ideal provides high nett housing density (say 24 dwellings per hectare), with an overall density which accommodates food provisioning and in situ waste management at a gross 12 dwellings per hectare). My 2004 *Sustainable Urban Planning* (Chapter 5) elaborates.

167 **The major challenge...** I have a friend now living solo in a three-bedroom dwelling which housed five people in the 1970s, eight people in the 1930s, and twelve people after the house was built in 1900. It is reported that solo-occupancy households now make up about half the Southern Californian urban household mix. 'For Transition Towns' the urban contract-and-converge movement, consult the Rob Hopkins *Transition Handbook* (2008). Wikipedia also gives well expressed information.

168 **In the matter of green building...** ESD — attributed to Norman Foster, probably via Paolo Soleri (ecotecture and arcology). An extensive and worthy treatment of sustainable building specifics is available from Johann Berhardt (editor) *A Deeper Shade of Green: Sustainable Urban Building and Architecture*, 2008.

169 **In terms of both...** At the UN climate change conventions in Bali (2007) Copenhagen (2009) and Cancún (2010), the impasse over North-South differences ended-up with an agreement to reduce emissions to half the Millennium Date level, but in a non-binding 'aspirational' way! The Durban 2011 convention made further quasi-binding progress.

170 **The alternative energy...** Concentrated Solar Power (CSP) is well understood

at a small scale, using silicon solar wafers for producing electrical energy for battery storage, and with hot water heat absorption devices. On a large scale, and much cheaper and longer lasting, are heat concentrating mirror farms, now a feature of Spanish and South Western US prototype projects.

171 **Throughout the habitable...** Here's an endnote to a 2012 Christmas Card from sunny Brisbane "We've had our solar power system hooked-up to the grid, and enjoy watching the wheel go round — backwards!"

172 **Outlaw fiscal buccaneering...** The reference to 'financial buccaneering' connects to the Enron, Bear Stearns, Lehman Brothers, and Merrill Lynch crises: lack of transparency (opacity), the use of off-balance sheet instruments (obfuscation), and blatant dodging of responsibility (dishonesty). Part of the solution may be found in adoption of the International Accounting Standards Committee's development of a single set of global accounting standards in conformity to transparent and comparable 'general purpose' statements; including universal adherence to XBRL (Extensible Business Reporting Language) usage.

## CHAPTER 14: TARGETS

173 **Throughout these pages...** Karl Popper, *The Open Society and Its Enemies*, 1945.

174 **The first of the three...** On the matter of 'supply meeting demand'. Immediately prior to the mid-2008 recession, oil prices reached $140 a barrel. A year later they had halved to $70; a clear example of supply meeting a recessionary falloff in demand with a price drop; and also an indicator that the minimum known profit that can be obtained from supplying oil is $70 a barrel! Free-to-water soluble nitrate disposal parallels the free-to-air carbon gas avoidance of externality costs.

175 **Plantation forestry...** Despite the fact that the Amazon, Congo and South Asian rainforests provide the optimum per hectare carbon gas sequestration, these natural carbon sinks are still being felled at an alarming rate for timber and conversion into charcoal. REDD (a United Nations scheme for Reducing Emissions from Deforestation in Developing countries) seeks to halt logging by initiating rights-payments to the tropical forest communities; and for those 'carbon credits' to be on-sold for their real worth and a tidy profit. REDD thus has its dubious side, styled as 'carbon piracy'. Overall the tropical rainforests suffer from 'cheat lie steal' mechanisms involving high-level and local corruption, along with obstacles to monitoring and enforcement.

176 **The $CO_2$ absorptive...** Pastoral farming in New Zealand generates half that nation's emissions damage as a consequence of the lethal methane and nitrous oxide 'flatulence' (mostly, it seems, burping) from sheep and cows. In total NZ's man-made greenhouse gas emissions amount to 0.2% of the world aggregate.

177 **Also important...** Fossil water aquifers in low-rainfall areas will not be replenished; so, after abstraction, that water ends-up as an additional minor

contribution to sea level rise.

178 **Mechanised feral fishing...** With oil costing around $140 a barrel by mid-2008, the much hammered marine ecosystem was getting a respite as it became profitless for the seafloor trawlers and tuna fleets to put to sea. That was good news for the biosphere above and below the sea surface. The enviro-political response to this situation should have been acceptance; instead ways were found by 'concerned governments' to subsidise fuel, make-over the fleets, and preserve the fisher's occupations as profiteering exploiter's of a resource common.

179 **This part of the discourse...** Not to be overlooked is the vast toxic release of methane from the now thawing tundra landscapes; along with the methane and nitrous oxide output of India's 290,000,000 cows and buffalo; and also along, of course, with pastoral animals worldwide.

180 **The symmetrical...** Throughout the Far East, fertilising the kitchen garden with human household waste is common practice, inducing Western aversion, but with little evidence of adverse health effects. In the West it is now common to return large amounts of sewage, processed into fertiliser concentrate, onto farmland, but with reports on the adverse effects to human health. The jury is out; my simple conjecture is that there is a difference between raw human household waste put directly onto the kitchen garden, and industrially processed human and other waste concentrating a cocktail of toxins and pathogens, which is spread over farmland in large concentrations.

181 **Worse, with the...** Heavy oils are dirtier and more difficult to refine relative to light petroleum oils. Oil sands and oil shales are even more challenging to refine, and in the process are environmentally damaging, producing three times as much greenhouse gas pollution as conventional oil production. Deposits are found in thirty countries, with North America (particularly Alberta in Canada), China, Mongolia and Russia holding major reserves.

182 **Here is a seven-way matrix...** Problem is, nuclear fusion (parodied as 'the fusion delusion'!), supplying energy with little intrinsically adverse environmental effect, would induce an onrush to population increases and reaccelerated consumer growth and waste discard. A multinational fusion project at Cadarche in southern France will take thirty years to construct and may come into production by 2026.

183 **When less fossil fuel...** 'Increasing prices' simply promotes a pointless longer-term revenue flow for oil producing cartels. For example Dubai, which fancies a future as a tourist attraction, is able to now invite high-end customer's to take their ease on under-cooled beach sands and dip into refrigerated swimming pools!

184 **A feature of building...** The simple stricture in relation to cement is to improve production efficiency, cut-down and cut-out its use, and enlist alternatives. While working on his doctorate (Cambridge circa 1980) John Coyne discovered that manufacturing the Standard British Brick required an energy input equivalent to the then average daily average British household all-up energy consumption.

185 **Industry absorbs around...** Two taxation proposals with considerable bite (of notable interest to the French administration) are Tobin's suggested Transaction Tax on all foreign currency dealings, and an Airline Ticket Levy.

186 **The worker input...** The 'lesser amount of work' statement could be all wrong; for in a less mechanised, yet well organised and locally grounded society, there would probably be an increased level of direct labour input to most activities.

187 **The Victoria Transport Institute...** Transport Demand Management strategies from the Victoria Transport Policy Institute (2002) are listed here in abridged and rearranged form:

| POLICY REFORMS | TRANSPORTATION OPTIONS |
|---|---|
| Access management | Alternative work schedules |
| Car-free planning | Cycle and pedestrian improvements |
| Commute trip reduction | Bike/transit integration |
| Major operator management | Car and ride sharing |
| Institutional reforms | Park and ride |
| Education transport planning | Flexitime |
| Special event management | Shuttle services |
| Regulatory reforms | Taxi service improvements |
| TDM and Marketing reforms | Transit improvements |
| Freight management | Design improvements |

| REDUCED DRIVING | LANDUSE MANAGEMENT |
|---|---|
| Walking and cycling | Bicycle parking |
| Financial incentives | Pedestrianised streets |
| Congestion pricing | Car free districts |
| High fuel taxes | Mixed land use |
| Reward high car occupancy | Efficient location |
| High parking prices | Parking solutions/management |
| Pay as you drive | Smart growth planning |
| Road pricing | Transit oriented development |
| Speed reduction | Optimally shared parking |
| Vehicle use restrictions | Clustered land uses |

In the wider context, convoy trains, trailer trucks, and stretched cargo ships are relatively pay-load efficient.

188 **Environmentally sound...** As with Norman Foster's Ecotecture, and the ZED *Zero Energy Development* (!) concept. Refer also to the previous Endnote.

189 **The unacceptable...** Further to 'childlessness' there is this from Jonathan Franzen author of *Freedom* (2010: interviewed by Ed Pilkington) "... given a world of rather striking finite resources, and a rather inflated and already

unsustainable population, it's an interesting thing to call a childless person selfish."

190 **The core real world...** The only, and at present merely investigative technological alternative to sequestering the excess mineral carbon uptake is Carbon Capture and Storage (CCS); which, to date, requires equivalent to 33% of the carbon consumed to drive the technology.

## CHAPTER 15: PART B: OVERVIEW & SUMMARY

191 **A humbling perspective...** Point 1: Professor Smith's full wording "Put simply and alarmingly, it means that if an economy grows, for example, at the modest rate of 3% a year, then in 23 years the activity (the size of the economy) will have doubled and in that doubling period we will need to use resources equal to all the resources consumed in the history of the economy." [The 95th Thomas Hawsksley Memorial Lecture, 2007]. Point 2: In 2007 (the last full year of all-round mega growth) the global economy grew by 5.4%. Point 3: a different angle on 'doubling' comes through from the Yann Arthus-Betrand DVD *Home* (2009) "In fifty years, in a single lifetime, the Earth has been more rapidly changed than by all previous generations."

192 **One hundred and fifty years ago...** J.S. Mill *On Liberty* (1859) in which he also reasoned that "It is only in [poor nations] that increased [rent seeking] production is still important.". JSMs reasoning was furthered by the lesser known Kropotkin (1889). One hundred years later the contemporary case for steady state economics has been advocated by many; most succinctly by Daly and Cobb, *For the Common Good* (1989).

193 **The gulf in time...** From (Mill 1859) *On Liberty* "...the sole end for which mankind are warranted, individually or collectively [is]... to prevent harm to others. His own good, either physical or moral, is not a sufficient warrant."

194 **Of course the World...** William Greider (1997) provides this flashback on my Chapter 7 reasoning '[The economic growth] machine has no internal governor to control the speed and direction. It is sustained by its own forward motion, guided mainly by its own appetites. And it is accelerating."

195 **The first climate change...** The eulogies-in-a-vacuum became quite wacky with the July 2008 G8 Canute-style communique. The catchy 20/20 through to 50/50 proclamation (carbon reductions of 20% by 2020 and 50% by 2050) was put out as an appeal for individual countries to adopt as their goal, voluntarily! And would that be 20% and 50% of 1990 Kyoto levels? Or 20% and 50% of Millennium levels? Or 20% and 50% of 2008 levels? By July 2009 the goals had morphed to 80% reduction (of 2009 levels?) by 2050, with no mention of curbs on the growth of population. And a final dig; this from the previous and also logic-free 2007 G8 communique which noted "... that the [then mid-2007] world economy [was] in good condition and growth evenly distributed"!

## APPENDIX A

196 **Her construct…** On 'freedom': the argument from the political 'right' is that the regulation of emissions and the like puts 'personal liberty at risk': a stance with its origin in the swashbuckling 'freedom' to produce-profit-pollute by big-oil, big-mining, big-tobacco, big-chemical and big-money.

www.ingramcontent.com/pod-product-compliance
Lightning Source LLC
Chambersburg PA
CBHW070746220326
41520CB00052B/905